MUSKEG ENGINEERING HANDBOOK

CANADIAN BUILDING SERIES

Sponsored by the Division of Building Research
National Research Council of Canada
Editor: Robert F. Legget

1 *Canada Builds*, by T. Ritchie

2 *Performance of Concrete,* edited by E. G. Swenson

3 *Muskeg Engineering Handbook*, edited by Ivan C. MacFarlane

Muskeg Engineering Handbook

By the Muskeg Subcommittee
of the NRC
Associate Committee on
Geotechnical Research

EDITED BY

Ivan C. MacFarlane

UNIVERSITY OF TORONTO PRESS

NATIONAL RESEARCH COUNCIL OF CANADA

ASSOCIATE COMMITTEE ON GEOTECHNICAL RESEARCH

CARL B. CRAWFORD, *Chairman*

MUSKEG SUBCOMMITTEE

DR. N. W. RADFORTH*	University of New Brunswick, Fredericton, NB. *Chairman*
C. O. BRAWNER*	Golder, Brawner and Associates, Limited, Vancouver, BC
C. T. ENRIGHT	Hydro-Electric Power Commission of Ontario, Toronto, Ontario
DR. R. M. HARDY	University of Alberta, Edmonton, Alberta
T. A. HARWOOD	Defence Research Board, Ottawa, Ontario
J. V. HEALY	Newfoundland Department of Agriculture, St. John's, Newfoundland
R. A. HEMSTOCK*	Imperial Oil Limited, Calgary, Alberta
I. C. MACFARLANE†	National Research Council of Canada, Ottawa, Ontario
DR. J. TERASMAE	Brock University, St. Catharines, Ontario
G. TESSIER*	Ministère de la Voirie du Québec, Québec
MISS J. BUTLER	(Secretary) National Research Council of Canada, Ottawa, Ontario

*Member, Muskeg Handbook Editorial Committee.
†Chairman, Muskeg Handbook Editorial Committee.

Contents

Foreword

Canada includes within its terrain at least 500,000 square miles of muskeg. Only in relatively recent years has this remarkable organic terrain been subject to scientific study. This study has developed in keeping with Canada's northern development and the need for more accurate prediction regarding transportation arrangements in undeveloped areas that are featured by muskeg.

The National Research Council of Canada, through its Associate Committee on Geotechnical Research (formerly the Associate Committee on Soil and Snow Mechanics), first gave serious consideration to the problem of muskeg in 1947. It was fortunate to obtain the interested co-operation of Dr. N. W. Radforth then of McMaster University, whose palaeobotanical research had direct relevance to the scientific study of muskeg. The Defence Research Board shared this interest. With financial assistance from both these bodies, muskeg studies in Canada had a slow but steady beginning. A turning point was reached when the first Canadian muskeg conference was held in 1955. As a result, national attention was focused on the problems associated with muskeg. The subsequent annual meetings have continued to be useful, and this Handbook is the outcome of the discussions that have taken place at them.

The Handbook presents a review of the state of the art of muskeg utilization, with particular emphasis on Canadian development. It has been written by experts in the various aspects of muskeg research and practice under the guidance and direction of the Muskeg Subcommittee of the Associate Committee on Geotechnical Research. Although the styles of the individual contributors have been preserved, the Handbook has been edited to ensure a minimum of duplication and a maximum of uniformity in terminology and the use of symbols. The practical approach has been emphasized and each chapter has an extensive bibliography to assist with further studies of particular facets of the subject.

The first three chapters of the Handbook reflect the range and extent of the research carried out by the pioneer in this field in Canada, Dr. N. W. Radforth. These chapters also show how the information from fundamental research can be put to practical engineering use. Chapter 4 is a compilation of available information on physical and chemical properties of peat relative to engineering design. Chapter

5 presents advice on preliminary investigations to be carried out in muskeg prior to any engineering operation, while Chapter 6 is a thorough treatment of the subject of road and railway construction in muskeg areas, perhaps the most common muskeg problem encountered by engineers. Chapter 7 discusses a wide variety of individual and important problems presented by muskeg, ranging from corrosion to construction of airstrips in organic terrain. Chapter 8 treats another very common muskeg problem in Canada, that of trafficability and vehicle mobility.

It is a pleasure to be able to include this Handbook in the Canadian Building Series which, in time, will form a collection of textbooks on the science and technology of building, using that word in its broadest sense, the sense so well indicated by the companion word in Canada's other language – bâtiment.

Division of Building Research   R. F. LEGGET
Ottawa, 1969   *Director*

Preface

The Muskeg Subcommittee is one of several national groups devoted to the task of interpreting Canada's terrain. It is an interdisciplinary organization composed of engineers and scientists, and reports to the Associate Committee on Geotechnical Research of the National Research Council of Canada.

The members of the Subcommittee and their associates have been active for a number of years in the promotion of technical sessions at various centres across Canada. At these sessions, known collectively as the Muskeg Conferences, accounts of muskeg research have been presented and co-ordinated in the form of Annual Proceedings. As a self-imposed challenge, the Subcommittee undertook to consolidate its work in handbook form under the editorship of Mr. Ivan MacFarlane, Research Adviser to the Subcommittee, in the belief that in this form the account would lend itself more readily to application.

Fredericton, 1969   N. W. RADFORTH

ACKNOWLEDGMENTS

Sincere appreciation is recorded for the efforts of all contributors to this Handbook, without whose willing co-operation its appearance would never have been possible. A word of thanks is also due to the members of the Editorial Committee for their assistance, and particularly to Mr. G. Tessier, who assisted in the onerous task of compiling the subject index.

The Editor is grateful for the continued interest of Miss F. G. Halpenny, Miss J. C. Jamieson, and Mrs. G. H. Stevenson, University of Toronto Press.

Special appreciation must also be expressed for the support, moral and otherwise, of this project on the part of Dr. R. F. Legget, Director, Division of Building Research, and past Chairman of the Associate Committee on Geotechnical Research.

Ottawa, 1969   IVAN C. MACFARLANE

CONTRIBUTORS

J. ADAMS   Supervising Engineer, Soils Section,
Structural Research Dept.,
Hydro-Electric Power Commission of Ontario, Toronto

C. O. BRAWNER   Golder, Brawner and Associates Ltd.,
Consulting Civil Engineers, Vancouver

D. CAMPBELL   Senior Construction Equipment Engineer,
Hydro-Electric Power Commission of Ontario, Toronto

C. T. ENRIGHT   Roads and Railroads Supervisor,
Hydraulic Development Dept.,
Hydro-Electric Power Commission of Ontario, Toronto

DR. R. M. HARDY   Dean, Faculty of Engineering,
University of Alberta, Edmonton

T. A. HARWOOD   Head, Geophysical Section, Defence Research Board,
Dept. of National Defence, Ottawa

R. A. HEMSTOCK   Arctic Co-ordinator,
Producing Dept., Western Regional Office,
Imperial Oil Limited, Calgary

DR. K. V. HELENELUND   Institute of Technology, Helsinki, Finland
(Visiting Scientist, Division of Building Research,
National Research Council of Canada, 1966–7)

W. J. KEYS   Senior Design Engineer, Imperial Oil Limited, Calgary

I. C. MACFARLANE   Geotechnical Section, Division of Building Research,
National Research Council of Canada, Ottawa

M. MARKOWSKY   Supervising Design Engineer,
Transmission and Distribution Dept.,
Hydro-Electric Power Commission of Ontario, Toronto

G. E. MCCLURE   Structural Design Engineer, Manitoba Hydro, Winnipeg

N. J. MCMURTRIE   Senior Design Engineer,
Transmission and Distribution Dept.,
Hydro-Electric Power Commission of Ontario, Toronto

W. R. NEWCOMBE   Dept. of Mechanical Engineering,
McMaster University, Hamilton

F. L. PECKOVER	Engineer of Geotechnical Services, Canadian National Railways, Montreal
J. R. RADFORTH	Development Engineer, Muskeg Research Institute, University of New Brunswick, Fredericton
DR. N. W. RADFORTH	Director, Muskeg Research Institute, University of New Brunswick, Fredericton
J. E. RYMES	J. E. Rymes Engineering Ltd., Calgary
G. SCHLOSSER	Producing Dept., Imperial Oil Limited, Edmonton
G. TESSIER	Service des Sols et Matériaux, Ministère de la Voirie du Québec, Québec
J. G. THOMSON	Manager, Defence Engineering, Chrysler Canada Ltd., Windsor
DR. S. THOMSON	Dept. of Civil Engineering, University of Alberta, Edmonton
G. P. WILLIAMS	Geotechnical Section, Division of Building Research, National Research Council of Canada, Ottawa

GLOSSARY OF TERMS

Aapa fen   See Fen, aapa

Aapa moor   See Fen, aapa

Absolute specific gravity   See Specific gravity, absolute

Acidity   See pH

Acidity, total   See Total soil acidity

Airform pattern   An arrangement of shapes, apparent at a particular altitude, which is characteristic for significant terrain entities and their spatial relationship and thus useful in the application of aerial interpretation

A-line   In a plasticity chart, the A-line represents the empirical boundary between typical inorganic clays which are generally above this line, and the plastic soils containing organic colloids which are below it

Amorphous-granular   A descriptive term applied to one of the primary macroscopic elements of peat which is granular in nature but to which no particular shape can be ascribed

Angle of internal friction   See Internal friction, angle of

Apiculoid   A descriptive term designating a 5000 foot airform pattern characterized by a fine-textured expanse, bearing minute projections

Ash content   The ash or mineral residue remaining after a peat sample has been ignited, expressed as a percentage of the dry weight; also known as ignition loss. Expressed as ratio of dry weight, it is known as ignition loss ratio

A-value   A pore pressure coefficient, measured experimentally in an undrained triaxial test

Axon   A well-preserved, non-woody, fossilized plant component of peat, consisting of tubular axis, system of leafy appendages, and with the cell structure clearly defined. The maximum outside diameter of the linear component of the axon (when wet) is 1 mm

Bearing capacity   The ability of a soil to support load without shear failure or excessive deformation within the soil mass

Bearing capacity, ultimate   The ultimate value of the average contact pressure, or stress, transmitted by the base of the footing of a foundation to the soil, causing the soil mass to rupture or fail in shear

Blanket bog   Also blanket mire. See Bog, blanket

Bog   An area of confined organic terrain, the limits of which are imposed by the physiography of the local mineral terrain. Differentiated from "muskeg" mainly in terms of area but often because variations in coverage, peat structure, and topography occur more frequently than in extensive areas of organic terrain

Bog, blanket   Equivalent of blanket mire, also "Terrainbedeckendmoore." Peat formation initiated in basins, drainage axes, and on all water partings where the drainage slope is not too great, the peat forming a blanket over all gently undulating terrain. In Ireland it includes a variety of peatland types, both ombrogenous (water supplied by precipitation) and soligenous (water supplied by high water table), occurring between 200 and 700 m. Variation in surface vegetation is related to climate (and perhaps supply of atmospheric nutrients), geology, topography, state of erosion, land use, etc.

Bog mire   Confined organic terrain; equivalent of bog

Bog, raised   Equivalent of raised mire and "Hochmoore." Peat development initiated in basins, peat growth producing a dome or cupola rising above the mineral ground water table. The classic case is the Baltic raised bog with ring lagg

Bog, spruce   See Spruce bog

Bog, string   See String bog

Category, cover   Also referred to as a Cover formula. A combination of two or three Class letters arranged in descending order of prominence of coverage classes as estimated by an observer at ground level. (Classes with an apparent representation of less than 25 per cent are excluded from a category or formula)

Category, peat   A descriptive term, one of a series of 17, applied to combinations of primary elements of peat structure. See amorphous-granular, fine-fibrous, and coarse-fibrous

Classification, cover   A subdivision of vegetal coverage based upon difference in properties such as woodiness vs. non-woodiness, stature, texture (where required), and growth habit

Coarse-fibrous   A descriptive term applied to one of the primary macroscopic elements of peat which may be woody or non-woody and has a diameter greater than 1 mm

Coefficient of compressibility   A stress–strain ratio of a soil. Numerically it is the slope of the void ratio – pressure curve from a consolidation test

Coefficient of consolidation   Obtained by plotting degree of consolidation against square root of time for any particular consolidation test

Coefficient of earth pressure at rest   Ratio of the horizontal effective stress to the vertical effective stress for a condition of zero lateral strain during consolidation

Coefficient of secondary compression   Is expressed as the shape of the settlement – log time plot divided by the thickness of the peat sample at the beginning of the long-term or straight-line stage; is also expressed in terms of change in void ratio over 1 cycle of the logarithmic scale on a plot of void ratio vs. log time

Compressibility, coefficient of   See Coefficient of compressibility

Compression index   The slope of the void ratio – log pressure curve from a consolidation test. The larger the compression index, the greater the compressibility of a soil

Compression, secondary   Also called "Secondary consolidation." Is compression of a viscous or plastic nature, its magnitude affected by temperature, the nature of the soil, and the rate of stress

Compression, secondary, coefficient of   See Coefficient of secondary compression

Conductive capacity   An index of the rate of heat flow through a medium because of an imposed temperature regime

Cone index   An index of the shearing resistance of the soil obtained with the cone penetrometer. The number, although usually considered dimensionless in trafficability studies, actually denotes pounds of force on the handle divided by the area of the cone base in square inches

Consistency limits   The liquid limit, plastic limit, and shrinkage limit of a clay or other cohesive soil; the standard way of describing clays; also known as Atterberg limits

Consolidation, coefficient of   See Coefficient of consolidation

Consolidation, primary   The gradual compression of a soil due to a weight acting on it, which occurs as water is squeezed out of the voids in the soil

Cumuloid   A descriptive term applied to a 5000-foot airform pattern characterized as a coarse textured expanse with lobed or fingerlike "islands" prominent; components shaped like cumulus clouds

Density, bulk   See Unit weight, natural

Density, dry   See Unit weight, dry

Dermatoid   A descriptive term applied to a 30,000-foot airform pattern characterized as chiefly featureless and plane: a simple covering lacking ornamentation (skinlike in the fundamental and literal sense)

Diffusivity, thermal   See Thermal diffusivity

Drawbar coefficient   Drawbar pull divided by gross vehicle weight

Drawbar pull   Also known as net traction. That portion of the gross traction which may be extracted through a drawbar. Alternatively, the net traction may be used to supply the extra thrust required for hill climbing. The numerical value varies with track slip

Drumlin   An elongated or oval hill of glacial drift

Earth pressure at rest, coefficient of   See Coefficient of earth pressure at rest

Esker   A long, narrow, often sinuous, ridge or mound of sand, gravel, and boulders deposited between ice walls by a stream flowing on, within, or beneath a stagnant glacier

External motion resistance   See Motion resistance, external

Fen   Muskeg (mire or peatland) consisting of organic terrain supplied with water previously in contact with mineral soil. If the area or areas are large this minerotrophic condition arises through contact with mineral subsoil rather than from surrounding mineral soil or rock slopes

Fen, aapa   Mixed, wide open (unconfined) muskeg mostly flat and minerotrophic; with irregular microtopography, e.g. airform patterns Vermiculoid I, II, and III; sometimes terraced; equivalent of aapa moor

Fen, horizontal   Permanently flooded organic terrain (minerotrophic condition obtains)

Fen, retention   Horizontal fen in which flood water is retained for as much as a whole year. May be covered by FI, EI, or BFI

Fen, sloping   Aapa fen in less flat topography often in conditions of high rainfall; equivalent of "Hangmoor" – peat formation in relation to spring

seepage lines and perched water tables, usually along the sides of small valleys or in the valley heads; equivalent of "Flachmoor" – peat formation in open basins with permanent inflow and outflow streams, and subject to the ground water flowing through the system

Fine-fibrous   A descriptive term applied to one of the primary macroscopic elements of peat which may be woody or non-woody and has a diameter less than 1 mm

Formula, cover   See Category, cover

Gross traction   See Traction, gross

Grouser   The part of the track shoe which projects outward from the plate and mechanically engages the soil. Grouser aggressiveness refers to the depth of engagement with the soil

Growth habit   A contributory distinguishing property of vegetal coverage used in conjunction with stature and woodiness vs. non-woodiness to determine cover class. A description of plant form and arrangement

Hardness, temporary   See Temporary hardness

Heat capacity   See Volumetric specific heat

Horizontal fen   See Fen, horizontal

Hummock   A microtopographical feature, includes Tussock. Has tufted top, usually vertical sides, occurring in patches, several to numerous

Ignition loss   See Ash content

Ignition loss ratio   See Ash content

Internal friction, angle of   Also angle of shearing resistance. Is the slope of the stress–strain plot for a particular soil

Internal motion resistance   See Motion resistance, internal

Intrusoid   A descriptive term applied to a 5000-foot airform pattern characterized as a coarse textured expanse caused by frequent interruptions of unrelated, widely separated mostly angular "islands"; interrupted

Kame   A short ridge, hill, or mound of

stratified drift deposited by glacier meltwater

Lagg   A natural ditch around the periphery of a confined muskeg, particularly of a raised bog

Latent heat   See Volumetric latent heat of fusion

Liquid limit   The percentage of moisture (based on oven-dry weight) at which a soil will just begin to flow and assume the characteristics of a liquid

Marbloid   A descriptive term applied to a 30,000-foot airform pattern showing a polished marble effect

Marsh   Low-lying tract of land with a high water table, usually covered with grass or sedgelike plants growing directly on mineral terrain

Moisture content   See Water content

Moraine   Deposit of glacial drift, generally in ridges, formed at the margin of a moving ice sheet

Motion resistance, external   The portion of the gross traction required to move the vehicle under its own power

Motion resistance, internal   The part of the engine power used in overcoming mechanical losses between the engine and the track

Motion resistance, towed   The drawbar pull which must be exerted to move the unpowered vehicle through the test media. Although the engine is disengaged at the clutch, mechanical losses at the sprocket and in the differential axle are included. Loosely interchangeable term with rolling resistance

Mound   A microtopographical feature with a rounded top, often elliptic or crescent shaped in plane view

Muskeg   The term designating organic terrain, the physical condition of which is governed by the structure of peat it contains, and its related mineral sublayer, considered in relation to topographic features and the surface vegetation with which the peat co-exists. (See Organic terrain)

Net traction   See Drawbar pull

Organic content   The weight of the or-

ganic material present in a sample of soil, expressed as a percentage of the dry weight

Organic terrain   A tract of country comprising a surficial layer of living vegetation and a sublayer of peat or fossilized plant detritus of any depth, existing in association with various hydrological conditions and underlying mineral formations

Peat   A component of organic terrain consisting of more or less fragmented remains of vegetable matter sequentially deposited and fossilized

Peat plateau, extensive   Also even peat plateau. A topographical feature usually extensive and involving sudden elevation

Peat plateau, irregular   A topographical feature often wooded, localized, and much "contorted"

Peat ridge   See Ridge, peat

Permafrost   A thermal condition of earth materials such as rock and soil when their temperature remains below $0°$ C continuously for a number of years, which may be as few as two or as many as tens of thousands

Permeability   The property of a substance which permits the passage of fluids through the pores of the material

pH   The measure of soil acidity. Defined as the negative logarithm of the hydrogen ion concentration in an aqueous suspension of the soil

Planoid   A descriptive term applied to a 5000-foot airform pattern characterized by an expanse lacking textural features, plane

Plastic limit   The lowest percentage of moisture at which a soil will cease to behave as a plastic mass and will begin to assume a condition of semi-solidity

Plasticity index   The numerical difference between the liquid and plastic limits; indicates the degree of plasticity of a soil

Polygoid   A descriptive term applied to a 5000-foot airform pattern characterized by a coarse-textured expanse cut by intersecting lines; bearing polygons

Polygon, free A topographic feature having many sides and forming a rimmed depression

Preconsolidation A construction technique whereby a load in excess of that which will be finally carried by the soft stratum is placed and allowed to settle until the ultimate settlement that would occur under the final load has been reached. The excess load or surcharge is then removed and the construction is completed; also called preloading

Primary consolidation See Consolidation, primary

Pull-slip test A method of measuring net traction as the track slip is increased in increments by the external application of drawbar restraint

Raised bog Also Raised mire. See Bog, raised

Resistivity, soil See Soil resistivity

Retention fen See Fen, retention

Reticuloid A descriptive term applied to a 30,000-foot airform pattern characterized by a network effect

Ridge, peat A topographic feature similar to mound, but extended, often irregular, and numerous; vegetation often coarser on one side

Rolling resistance See Motion resistance, towed

Secondary consolidation See Compression, secondary

Sensitivity A term designating the sensitivity of a cohesive soil to remoulding. It is the ratio of the undisturbed shear strength to the remoulded shear strength

Sloping fen See Fen, sloping

Soil resistivity The electrical resistance of a soil. It is usually measured as the number of ohms resistance across one cubic centimetre of soil (ohm-cm)

Specific gravity, absolute The ratio of the density of soil solids (in oven-dried condition) to that of distilled water

Specific gravity of soil solids (oven-dried condition) See Specific gravity, absolute

Spruce bog A term in common use loosely applied to confined areas of organic terrain where coniferous trees (often not spruce) are a prominent feature of the vegetal coverage

Stipploid A descriptive term applied to a 30,000-foot airform pattern. The stipploid condition seems to be constructed of closely applied dots

String bog See Vermiculoid III

Swamp Similar to marsh but usually with higher water table and interruptions in the vegetal mat

Temporary hardness The quality of water containing dissolved salts of calcium and magnesium, i.e., bicarbonates which are removable by boiling

Terrazoid A descriptive term applied to a 30,000-foot airform pattern that shows a "patchwork" quality

Thermal conductivity An empirical coefficient defined as the number of calories per second flowing through a plate 1 cm square and 1 cm thick with a temperature difference of $1°$ C between the two surfaces of the plate

Thermal diffusivity An index of the facility with which a substance will undergo a temperature change

Thermokarst Subsidence of the ground surface in permafrost regions, producing uneven undulations and hollows caused by the melting of ground ice

Total soil acidity The total quantity of ionizable hydrogen in the soil

Towed motion resistance See Motion resistance, towed

Track pitch Originally the centre-to-centre distance between track shoe pins. On a continuously flexible track this is the distance between grousers

Traction, gross The total thrust developed by the vehicle tracks. Equivalent to the track belt tension

Traction, net   See Drawbar pull

Tussock   See Hummock

Type, peat   See Category, peat

Unit weight, dry   The weight of dry material (oven-dried to a constant weight) in a unit volume of original material

Unit weight, natural   The total weight (including solid material and any contained water) of a material per unit volume (including voids)

Vehicle cone index   The minimum cone index that will permit the vehicle to complete 50 passes

Vermiculoid, primary   A descriptive term applied to a 5000-foot airform pattern characterized as striated, mostly coarse-textured expanse; featured markings tortuous. Subdivided into three secondary configurations, Vermiculoid i, ii, and iii

Vermiculoid i   As above, but striations webbed into a close net and usually joined

Vermiculoid ii   As above, but striations in close association, often foreshortened, and rarely completely joined

Vermiculoid iii   As above, but striations webbed into an open net, usually joined, but very tortuous

Vertical snaking   The vertical undulations of the track on elastic ground caused by the wheels running over it. Horizontal snaking can occur during turns and side hill operation

Void ratio   The ratio of the volume of voids to the volume of soil solids

Volumetric latent heat of fusion   The number of calories required to change the ice in one cubic centimetre of soil from solid to liquid without temperature change

Volumetric specific heat   The number of calories required to raise one cubic centimetre of soil $1°$ C

Water content   The amount of water contained in the voids of a soil. In standard soil mechanics terminology it represents the loss in weight expressed as a percentage of the oven-dry material when the soil is dried to a constant weight at $105–110°$ C

SYMBOLS

a — thermal diffusivity; cross-sectional area of narrow-bore tube of permeameter

a_v — coefficient of compressibility

A — cross-sectional area of peat sample; area of load contact surface

A_c — ash content

B — width of strip load

c — perimeter shear coefficient

c' — effective cohesion intercept

c_u — apparent cohesion

c_v — coefficient of consolidation

C_a — specific heat of air

C_c — compression index

C_s — coefficient of secondary compression; specific heat of soil

C_v — volumetric specific heat

C_w — specific heat of water

σ — normal stress in soil element

σ' — effective normal stress

σ_t — tensile strength

e — void ratio

E — Young's Modulus

ϵ — angular strain

ϕ_u — apparent angle of internal friction

ϕ' — effective angle of internal friction

G — specific gravity of soil solids

γ — bulk density

γ_d — dry density

γ_w — unit weight of water

h — height of peat sample (beam tests)

H — thickness of peat layer (or sample) at beginning of secondary compression phase; height of shear vane

ΔH — change in thickness of sample (or layer) during primary consolidation phase; settlement

H_o — initial peat thickness

i — exponential parameter

k — permeability coefficient

K — thermal conductivity

K_o — coefficient of earth pressure at rest

l — length of peat sample (beam tests)

L — volumetric latent heat of fusion

ΣM_u — ultimate resisting moment

p — uniformly distributed load

Δp — consolidating load increment

p_c — critical load

p_u — ultimate bearing capacity

P — perimeter of loaded area

r — radius of load contact area; radius of shear vane

Σs — sum of shearing resistance in surface of rupture

S_e — elastic settlement

S_o — primary compression (consolidation)

S_s — secondary compression

S_t — total settlement

t — time interval of secondary compression

t_o — time of primary consolidation

T — vane torque

τ — shear stress

τ_{av} — average shear stress

τ_f — shear strength

τ_{max} — maximum shear stress

v — Poisson's Ratio

X_a — volume fraction of air in a soil

X_s — volid fraction of solid material in a soil

X_w — volume fraction of water in a soil

MUSKEG ENGINEERING HANDBOOK

1 Muskeg as an Engineering Problem

Chapter Co-ordinator
N. W. RADFORTH

1.1 THE MEANING AND SIGNIFICANCE OF MUSKEG

In North America the expression "muskeg" is in wide use. Those involved in the development of terrain are the primary users of the expression. The Chippewa Indians were among the first to recognize that the terrain to which "muskeg" is applicable has characteristic features. They named it "maskeg," which means grassy bog, a typical example of which is shown in Figure 1.1. This literal

FIGURE 1.1   Muskeg derived from Chippewa Indian "maskeg," meaning grassy bog

interpretation is not broad enough in implication to meet contemporary need. Modern argument insists that vegetation other than grass (for example, Fig. 1.2) very frequently occurs and "bog" has uncertain meaning for the uninitiated.

A survey to obtain the consensus of understanding on the accepted meaning of muskeg was started in 1945 by N. W. Radforth, with the encouragement of the National Research Council of Canada. It was found that several basic aspects of meaning were needed to satisfy most of those who used the expression. To provide widest satisfaction the following definition, which emphasizes "organic terrain" as a substitute expression, was provided: " 'muskeg' has become the term designating organic terrain, the physical condition of which is governed by the structure of the peat it contains, and its related mineral sub-layer, considered in relation to topographic features and the surface vegetation with which the peat co-exists."

"Muskeg" is thus a term commonly used to convey "organic terrain" (Radforth 1952). The definition recognizes physiographic character, vegetal cover, peat, environmental association with underlying mineral soil, and topographic character.

The living vegetation covering the terrain is composed of mosses, sometimes lichens, sedges and/or grasses, with or without tree and shrub growth. Usually combinations of these plant forms are found. Underneath this cover there is a

FIGURE 1.2   Contemporary usage of muskeg has broader connotation respecting vegetation

(a)

(b)

FIGURE 1.3   Peat (sometimes muck)
is constituted of partially fragmented
and chemically modified fossilized
plant constituents

FIGURE 1.4   Muskeg implies organic landscape in a three-dimensional sense

mixture of fragmented organic material derived from past vegetation but now chemically changed and fossilized (Figs. 1.3 (a), (b)). This material is commonly known as "peat" or (less acceptably) as "muck."

It is known that this subsurface material is highly compressible (MacFarlane 1958) compared with most mineral soils. The general terrain condition is typically very high in water content and is usually of extremely low bearing capacity. Implicit in the definition is a certain other connotation. The importance of three dimensions (area and depth) is implied (Fig. 1.4). The recognition that the materials (living cover and fossilized remains of past vegetation, i.e. peat) have generic association with the mineral base also has significance (cf. Fig. 1.5(a), (b)).

Other features appear implicit in an understanding of muskeg. The presence of water is implied because of its necessary role as a fossilizing agent in the formation of the peat. Additional inference entails the presence of aerobic conditions, usually (though not always) high acidity, humic acids, peculiarities in ionic relations, a partly colloidal constitution, specialized physical and mechanical properties imparted by the organic constituents, etc.

FIGURE 1.5   Muskeg signifies relationship between organic overburden and underlying mineral foundation

(a)

(b)

FIGURE 1.6   Special engineering treatment is required for access and construction in muskeg. Mining, reclamation, conservation of water, forestry, and agricultural application are examples of operations that require special design

For some people "muskeg" may have a broader connotation, for example, specialized environmental control. Others may think of the terrain in terms of its behavioural characteristics; for example, its response to the forces of erosion, its change in organization with modification in water deployment, its response to freezing temperatures, or characteristics of either the peat or the terrain in relation to insulation. Thus, the meaning of muskeg is complex and the phenomenon it represents should not be viewed in simple terms if application is anticipated. Emphasis on meaning teaches the user to understand at the outset that organic terrain contrasts markedly with mineral terrain and that evaluation of it must involve new methods of approach and research.

The importance of muskeg may now be appreciated. Engineers who must deal with it are already accustomed to designing for features characterizing mineral rather than organic terrain. The peculiarities of muskeg in terms of its properties arise embarrassingly in land-use engineering because these properties usually impose limitations on the application of standard design approach which is devised for ordinary terrain conditions. In the first instance, access to organic terrain, whether by foot or vehicle, is difficult and in some circumstances impossible. The significance of muskeg is such that special engineering treatment is required for access and construction. Special design is needed for mining, reclamation, water conservation, forestry and agricultural operations (cf. Figs. 1.6(a)–(f)). Establishment of primary and local road systems also calls for a particular approach where muskeg is encountered. Changes in terrain properties, for instance those associated with seasonal succession, further complicate the access problem (see Fig. 1.7(a), (b) showing vehicular access on the same kind of muskeg but at different seasons). The construction of foundations for a muskeg environment requires new design formulation and technique.

Operations such as exploration for and production of petroleum and other resources, including peat itself, require special consideration as does drainage control. Note, in addition, the examples suggested pictorially: (1) Figure 1.8, summer access over muskeg in a condition common to forestry practice; (2) Figure 1.9, establishing a pipeline in muskeg (note the immobilized dragline); (3) Figure 1.10, reclaimed muskeg. Where organic terrain is developed for use, engineers are invariably required and may be called upon to re-establish or intensify drainage. There must be awareness and special ability concerning the use of machinery for working the peatland, disposing of the peat, or reclaiming the land for agricultural purposes.

1.2 RECOGNITION OF MUSKEG CONDITIONS

Observers have little difficulty in recognizing organic terrain from the deck of an off-road vehicle or from a road. This is usually the case whether or not one can see exposed peat in the area. The reason is that where organic terrain commences

(a)

(b)

FIGURE 1.7   Vehicular access on a given kind of muskeg, but at different seasons

FIGURE 1.8   Summer access over muskeg in a condition common to forestry practice

FIGURE 1.9   Establishing a pipeline in muskeg

FIGURE 1.10   Reclaimed muskeg

there is a change in the type of vegetation, and the new type persists far enough in the field of vision to form a significant component of the landscape. On adjacent mineral terrain one can see, by contrast, great variability in the stature and form of individual plants and groups of plants constituting the cover.

On organic terrain, the stature of the vegetation resolves itself into characteristic orders of height, with a single order featured by neighbouring plants *en masse*. If

(a)

(b)

(c)

FIGURE 1.11   Vegetal cover on muskeg is regularized as to stature, form, and texture (woodiness or non-woodiness). Differentiation of cover results in the presence of areas of contrasting physical arrangements

there is a change in stature, it is more often abrupt than gradual. In addition to uniform stature of the constituent plants, form also becomes regularized (cf. Figs. 1.11(a)–(c)). Organic terrain may, and frequently does, show evidence of demar-

FIGURE 1.12   Building on a main street in Prince Rupert, BC, said to have one corner resting on a pile driven into peat over 70 feet deep

cation based on texture (woodiness as opposed to non-woodiness) as a feature conveyed by the plants *en masse*. Thus height, form, and texture in the cover, when considered comprehensively, serve not only to differentiate organic from mineral terrain, but also to enable the observer to distinguish different kinds of landscape within the organic terrain. The presence of areas of contrasting physical arrangements is often encountered, even in small expanses of organic terrain extending for only one or two acres.

On the organic terrain, when superficial vegetation is removed, there is no difficulty in recognizing the underlying fossilized organic matter. This is the peat. It varies in colour from straw through golden brown to chocolate and black. It is the "legacy" of past cover. It may be coarse or fine, depending upon the structure of the proportionate components of past cover that became fossilized successionally. Sometimes it is only 2 or 3 inches (5.1 or 7.6 cm) in depth, but there are cases where deposits 90 feet (27.5 m) deep are said to have been encountered (Fig. 1.12). The peat may be so dry as to show no evidence of free water even when a sample is squeezed, or, on the other hand, it may give up sufficient water to run between the fingers (Figs. 1.13–1.15).

FIGURE 1.13   Peat with ice incorporated

FIGURE 1.14.   Peat with no free water

FIGURE 1.15   Peat *in situ* showing presence of free water

FIGURE 1.16   Aerial photograph showing organic terrain filling the depressions between the folds of Precambrian igneous foundation

If the peat is shallow, there will be no difficulty in detecting the underlying mineral soil or rock. Radforth (1962a) has found a complete range of mineral soil types beneath peat of one kind or another. Pure clay, silt, sand, coarse gravel, or boulders may occur, but usually there are admixtures of mineral soil. The soils may have been derived from glacial or lacustrine activity or may be on eroded non-glaciated upland. Occasionally one may find muskeg lying almost directly on sedimentary or igneous rock (Fig. 1.16).

Differences contrasting vegetal cover for organic and mineral terrain are accompanied by a change in topography. This factor, apart from those characterizing the structure in vegetal cover, facilitates recognition of the immediate boundary of the muskeg. Topographic difference applies particularly and is most useful when the muskeg is "confined," that is, enclosed by upland of mineral soil or rock.

One may encounter organic terrain on slopes (glaciated or unglaciated), in deltas, or on foundations of ancient lakes and rivers on any continent in tropical, subtropical, temperate, or arctic zones. Sometimes it is associated with kettle-holes or other postglacial features of the land, for example, with kames, eskers, and drumlins (Figs. 1.17(a), (b) and 1.18).

When the peat has originated in a shallow basin, often the edge of the original depression in the landscape is overgrown by extensions of the peat deposit. Where this occurs, it is now known that the condition is detectable by the vegetal cover. It is not always a simple matter to discover the ultimate edge of the area of organic terrain under investigation. If the original peat deposit has spread excessively or a number of secondary centres each overgrowing have arisen spontaneously, analysis of the situation becomes very complex. This complexity will be appreciated if reference is to be made to the type of underlying mineral soil or to the prevalence of one type of peat or another in the organic overburden. In these circumstances, the organic terrain is "continuous" or "unconfined" (Radforth 1962b). The "interruptions" of features of inorganic landscape enclosing "confined" muskeg are lacking.

Where muskeg is unconfined, the vegetation will continue to show structural differentials, but often the differences noted characterize phenomena inherent in the organic overburden itself and are thus recent in origin. In both confined and unconfined muskeg, topographic features often characterize local conditions. These are known as microtopographic features and are sometimes helpful as indices of ready reference in recognizing local conditions within organic terrain (Fig. 1.19(a), (b)) (Radforth 1952, 1955).

1.3   ENGINEERING SIGNIFICANCE OF DISTRIBUTIONAL PHENOMENA

Concern for the muskeg phenomenon for engineering purposes will be broad or narrow, depending upon the kind of implication facing the observer or investigator. Ultimately, engineers must assess the muskeg state and the problems it presents

(a)

(b)

FIGURE 1.17   Muskeg boundary in apposition with kames (confined muskeg)

FIGURE 1.18   Muskeg boundary
in apposition with esker (confined
muskeg)

prior to a design intended to meet the specified objectives associated with one or
more operations. "Design" is the important word in this complex.

It is also important to recognize that muskeg with its panorama of characteristics
and peculiarities presents limitations. Often the limitations are characteristic,
and represent recurring circumstances. This is probably due to repetitious
states which for each type of muskeg produce a peculiar range of mechanical
potential. The behaviour of the design model will always give the same response
when the model is applied in such circumstances. This effect should reassure rather
than dismay the engineer because it helps in the delineation of safety factors and
may also facilitate prediction.

Another element in gaining adequate application is implied in the expression
"operations" and this circumstance imposes another kind of limitation for engineering implementation. Usually the terms of operations are inflexible. Time, method,
and financing are involved. From the point of view of engineering interests in
muskeg, therefore, there are three areas to which attention must be directed with
some emphasis: (1) design and the two limiting factors, (2) total muskeg constitution, which is unique and fixed, and (3) operations in which stringency is inherent. It matters little what branch of engineering is involved; planning, design,
and implementation must adjust to a formula in which conditions imposed by the
limiting factors are satisfied. This circumstance specifies for and controls engineering practice in muskeg. If the designer acknowledges it, empiricism will be avoided
and testing will be minimized; much preliminary analysis will be unnecessary and
avoided.

Inadequacy or failure will ensue, or over-design will arise if adjustments for
proposed installations do not allow for total environmental assessment of the
organic medium. If the operation contemplated is highway construction, the
design engineer must be aware of the dynamics of the water regime in the muskeg

(a)

(b)

FIGURE 1.19   Contrast in microtopographic difference in two areas of organic terrain

FIGURE 1.20   One condition of muskeg which invariably floods in spring

to be crossed. Design should not be regarded as complete until he has considered
the seasonal behaviour of the water–ice components within the environment. For
instance, one kind of muskeg will invariably flood in spring, another will not
(Fig. 1.20). Design for consolidation and stability, construction of grade and
subgrade, and location of culverts will require appropriate treatment. Requirements for intensity and frequency of loading will also have to be considered in
relation to total, year-round environment. The process of modification naturally
commences with the survey for road location and if, finally, the muskeg must be
crossed, its fixed conditions become limiting and must be met.

The mechanical engineer is also closely involved with the problems posed
by muskeg. The importance of specified design and limitation already mentioned
is encountered in vehicle mechanics when off-the-road access over muskeg is
required. There are examples where expensive vehicles and drainage machinery
have been developed without proper consideration of the inherent, fixed design in
the organic terrain. Immobilization, when it ensues in these cases, portrays failure
of purpose, serious interruption in operations, and perhaps more significantly
inadequacy, if not irresponsibility, in engineering practice.

It is not generally known that the feature of muskeg which now provides the

greatest difficulty for amphibious vehicles is microtopography. One would expect
that shear or bearing capacity might provide the major source of trouble. This
was the case several years ago perhaps, but it is no longer so. In vehicle design, the
limits of tolerance for ground pressure are known, whereas much still has to be
learned about design for suspension systems (Fig. 1.21 (a), (b)). This is not to
say that the other features characterizing muskeg, now sometimes expressed in
quantitative or quasi-quantitative terms, are not significant, but they are not as
important as they once were. Ultimately, for optimum success, all repetitive
features of the terrain must be matched by analogous mechanical equivalents aptly
co-ordinated in the vehicle.

The muskeg environment calls for specialization in sampling. The design of
testing equipment requires interdisciplinary application and involves the co-
operation of electrical, mechanical, hydraulics, and materials engineers with soil
and design experts. The muskeg environment also requires the judgment of the
geologist, geographer, and applied palaeobotanist whose interests overlap those
of engineers. The agricultural and forestry experts working with the engineers
provide specialized knowledge without which operational success cannot be
expected.

Chemical engineering in muskeg studies also involves specialization. The manufacture of commodities from peat is virtually undeveloped, but research must
precede utilization. It is here that there is need for the services of chemical engineers. The physical and chemical differences in peat which could lead
to its exploitation have not yet received sufficient attention. For example, there
is interest in the use of peat in the pelletizing process in iron ore refinement. In
this connection, insufficient attention has been given to the primary constituents
of peat and to the properties of various peat structural categories. These may vary
widely, but in a reasonably predictable way.

Table 1.1 categorizes engineering discipline and the muskeg environment in
relation to engineering significance.

1.4   DESIGN FOR RESEARCH

Confusion sometimes arises between engineering research and engineering testing.
It is important to differentiate between these kinds of endeavour as a principle
in a design system. Examination of Table 1.1 suggests, with respect to all the items
listed, that interpretative study reflecting organization in muskeg establishes the
main direction to be taken in research. In an effort to reveal the unknown – the
basic objective of research – one could easily slip into an elaborate programme of
ad hoc field sampling and testing to secure data for an immediate engineering need.
Economic and operational pressures often encourage this approach. The results,
having been empirically sought, are too often narrow in their application and

(a)

(b)

TABLE 1.1
Engineering significance of muskeg

Engineering discipline	Purpose of involvement	Single examples
Aerial interpretation	Road location	Service road (Fig. 1.22)
	Drainage survey	Peat exploitation
	Reservoir location	To satisfy requirements for dam site
	Communications installations	Rail-bed maintenance
	Hydro-electric systems locations	Service transmission line right-of-way
	Land utilization interpretation	Economics for timber exploitation
	Foundations analysis	Tank farm installation
	Resource location	Drilling sites (petroleum)
Design engineering	Operations optimization	Military transport
	Transportation research	Track design
	Trafficability analysis	Off-road supply
	Vehicle-terrain adaptation	Specified off-road access
	Mechanical design (appliances)	Drainage ploughs
	Instrumentation	Peat density probe
Hydraulics	Drainage assessment	Watershed capacity
	Drainage control	Dam construction
	Drainage manipulation (forestry, agriculture, highways, general construction)	Water diversion techniques
	Peat exploitation	Pipeline construction
	Peat product manufacturing	Water content assessment
Agricultural engineering	Vehicle adaptation	Four-wheel-drive tractor
	Equipment adaptation	Ditch location
Engineering survey	Townsite location	Planning appropriate road network
	Services distribution	Positioning sewage
	Subsurface ice evaluation	Foundation stabilization
Sanitation engineering	Industrial waste dispersal	Sewer location
	Water utilization	Water source selection
	Pollution control	Peat water diversion
	Sewage disposal	Disposal bed improvement
Engineering research	Materials evaluation	Bearing strength
	Physical properties of peat	Permeability study
	Peat mechanics	Stability analysis
	Muskeg characterization	Peat structure representation
	Prediction for utilization	Pelletizing
	Environmental interpretation	Range of foundation conditions
	Optimization studies	Pulpwood operations
	Environmental manipulation	Water diversion

FIGURE 1.21   An amphibious vehicle which can successfully accommodate to bearing capacity and strength of muskeg, but not to microtopography, if its performance is to be regarded in relation to the optimal requirements

FIGURE 1.22   Service road through tree-covered muskeg area (Hydro-Electric Power Commission of Ontario, Little Long Rapids Development)

inspire little confidence because they fail to interpret the total muskeg environment. They signify testing but not research. They do not portray fact in an adequate context.

As an example, it is not enough to embark upon an elaborate scheme for determining the properties of peat and associated mineral soil in anticipation of road design since seasonal change in the water regime, for instance, will so change local conditions as to seriously influence subsidence. Cases exist for which a soil sampling programme (or something less) has been accepted. Usually it has been accepted on an assumption that basic design for roads has been established. The sampling for properties, it should be emphasized, constitutes testing and is necessary, but the results will not guarantee stability of the road until the proposed engineered configuration meets all the implications of the muskeg factor (Fig. 1.23). In addition to properties of material, several other factors should be assessed: affect of physiographic dynamics, disposition of peat types, association with mineral soil (interface geometry and its causes), behaviour of environmental water, and stability of the site. This does not exclude the possibility of revising basic design to accomplish new levels of standardization. Formulation of design for categories of cases still constitutes practical procedure, but a fair measure of sophistication is a necessary ingredient.

FIGURE 1.23   An example of severe subsidence in a highway over muskeg

Theory and hypotheses provide the background for interpretative judgment on muskeg conditions. There is an argument that propositional analysis imposes frustration on engineering practice because engineering application follows from factual not hypothetical evaluation. Theory and hypothesis about muskeg conditions still beg the question in many instances when engineering requirement is imposed. On the other hand, neither theory nor hypothesis will survive unless they arise from observation involving data derived by measurement. They set the pattern of evaluation and assist in departmentalizing reliable approach and field data to good practical purpose. Thus, where a road is to be constructed in the North, it is unwise to ignore the hypothesis that permafrost exists beneath peat cover of a certain definable type. That it does exist is so probable (and reasonable) that it has been demonstrated and tested as a predictable circumstance. Where it exists spasmodically, a road foundation, particularly if installed in the late summer, will show differential subsidence if the design is inappropriate. There is no excuse for ignoring this possible muskeg–permafrost relation, even though proof of its existence is not statistically a fact.

If theory and hypothesis are accepted attitudes for engineering evaluation and design, the second argument is that engineering practice follows naturally on the disclosure of scientific principle and conclusion. To use examples so sharply specified as road subsidence or instability to illustrate the relation between research and testing may be considered irrelevant. It is believed, however, that engineering research is a matter for consideration at all levels of application and that, where muskeg is involved, every engineered event should be preceded by pertinent preliminary research. The event also involves special problems which must be solved before design consistent with good engineering practice can be judged as reliable. These are the problems that embody the research at whatever level they impinge.

1.5 MUSKEG AND ECONOMICS

Operations in which muskeg is a primary factor are governed not only by the physical limitations of the terrain but also by economic systems. In relatively undeveloped countries the economics usually centre on the problems of undisturbed terrain. When the terrain already has been exploited, secondary adjustments are evident in the physical properties of the peat. The changes are caused either by long-term drainage or manipulation of gravitational water or by fires. The effects of these processes on the terrain result from the consolidation of the peat and interruption of the homogeneity of the deposits. The secondary disturbances reduce the ordinarily different terrain examples to a single order of likeness. The effect is superficial and, because of this, the application of economics and the classification of the deposits in depth is difficult to achieve by reference to the surface cover. This is transient and secondary and therefore not truly representative of conditions that would have ensued had man not interfered.

Whether secondary manifestations are present or not, the first requirement for development or operations of all kinds is effective access to the area in question. At first, if there has been no secondary change in the terrain, vehicles with a ground pressure no greater than 1.5 pounds per square inch (0.105 kg per sq cm) should be used to gain access in order to assist in the initiation of a drainage programme. Where signs of secondary change exist, local consolidation in the peat sometimes permits the use of vehicles with a ground pressure of up to 5 pounds per square inch (0.39 kg per sq cm), provided the tracks are not aggressive to the point of destroying the top zone of the terrain profile where living and fossil components are mixed. Where open water occurs, vehicles should be amphibious. Observations gleaned from an adequate survey of the pertinent terrain conditions form the foundation for success in the economics of development.

Once effective access and optimum drainage have been achieved, the peculiarities of the imposed industry dictate the economics. In Britain, the Scandinavian countries, and particularly in Canada, the forestry industry relies upon year-round field activity. The economics of exploitation for organic terrain must be worked out accordingly. Special costing involving summer and interseasonal use of labour and the use of access and servicing roads for all phases of forestry operations have to be considered and absorbed in preparing cost estimates. Thus, the year-round

FIGURE 1.24   View showing peat stacks (blue peat), for use by the Highland Park Distillery, on the Orkney Islands

practice of forestry, either on or across muskeg, requires special economic evaluation entailing examination of special machinery – track and/or wheel vehicles (especially articulated ones) – to effect access, grubbing, road construction over muskeg, cropping, and transport in and over exploited muskeg areas. This is now being achieved with remarkable success in forestry operations in Canada and to some extent in Sweden. In Britain (for example, in Wales, northern England, and Scotland) drainage of organic terrain for controlled reforestation has become highly successful as an economic venture.

It has been pointed out (Hemstock 1959) that "wildcat wells in muskeg areas have shown summer road building and transportation costs as high as $500,000 per location." This statement was made in the hope of encouraging attention to the problems of working in muskeg. Hemstock has pointed to the need for scheduling work in the field to fit the seasons so that the work to be executed is of a type appropriate to the seasonal character of the muskeg in question. Emphasis is placed on the need for good engineering planning involving wisely chosen access routes and the use of airphoto interpretation and field reconnaissance. An approach to relieve economic pressure at the drilling stage involves a method called "island drilling." If muskeg forms part of the overburden, the drilling site is placed on an "island" of clay which is floated on the muskeg. The ensuing directional drilling increases the costs but lessens the engineering difficulty.

In the exploitation of the Pembina field, about $30 million was judged to be the price of dealing with the muskeg factor. Building gravel access roads over

FIGURE 1.25   Cracking of house foundation in Prince Rupert, BC, caused by differential settlement

muskeg costs about twice as much as it would if only mineral terrain were involved. Walsh (1957) indicated that access roads over muskeg in oilfields cost 1.45 times as much as those placed directly on mineral soil.

In northern development to assist in the exploitation of resources and to facilitate tourism, the Alaska Highway provides a useful communication route. Muskeg impeded development of this highway and the high costs of construction bear witness to this fact (US Government Printing Office 1961). During its construction, costs were kept to a minimum by avoiding either excavation of peat or placing of corduroy. Usually fill was placed directly on the muskeg surface. Costs to restore the grade after expected settlement amounted to about $15,000 per mile in five years (Thomson 1957). Where muskeg intervened, costs for grading were increased by 60 per cent. Thomson (1957) stated that the life of a road constructed over muskeg is shorter than it is for one where peat is lacking in the foundation. Also, the subsequent cost of patching and raising the grade has to be anticipated.

Davis (1961) and Harrison (1955) have both reported on the high and variable costs for roads over muskeg in establishing access and haul roads in forestry practice.

The principles of economics turn about certain special aspects of muskeg development. In government surveys the distribution and classification of muskeg are under consideration with a view to making the wisest use of the resource. On the northern fringe of land development in Canada, the existing economic principle appears to stress the use of muskeg for forestry purposes except in northern Alberta and British Columbia, where some agricultural use of muskeg has been suggested. As mining of peat becomes more economical, it is anticipated that a trend already begun to move large quantities of peat to the greenhouses of the south to increase crop productivity will be intensified. Recently, the mining of peat by hydraulic methods and its subsequent delivery by pipeline has been attempted in Canada. Because peat regenerates so slowly, it is classifiable as a non-renewable resource. Exploitation of it as an energy source (for example, in the Falkland Islands, Ireland, and the USSR) or as a product for use in industry (see, for example, Fig. 1.24) will require restudy if the outlook for the economics of food productivity continue to darken.

The main economic factor for organic terrain, pending further exploitation for certain major industries, relates to construction. The building of roads, tank farms, northern townsites and buildings is the contemporary concern (Fig. 1.25). In permafrost country the need to preserve the muskeg in its natural state is paramount in the construction of building foundations. The problem is how to maintain the insulating effect of the peat and vegetal cover to discourage differential and local thawing below the ground surface.

A most important economic implication of muskeg is related to the northern source of Canada's water supply. Modification of peat type following engineering activities undoubtedly alters the permeability of peat and may destroy or greatly change the natural control system for retaining seasonal gravitational water.

REFERENCES

DAVIS, A. E. 1961. Road construction for forestry practice in Northern Ontario. Proc. Seventh Muskeg Res. Conf., NRC, ACSSM Tech. Memo. 71, pp. 92–102.

HARRISON, W. C. 1955. Primary and secondary access over muskeg in forestry practice. Proc. Western Muskeg Res. Meeting, NRC, ACSSM Tech. Memo. 38, pp. 6–13.

HEMSTOCK, R. A. 1959. Muskeg – a review of engineering progress. Can. Surveyor, Vol. 14, No. 10, pp. 470–477.

MACFARLANE, I. C. 1958. Guide to a field description of muskeg (based on the Radforth Classification System), NRC, ACSSM Tech. Memo. 44, rev. ed. 36 pp.

RADFORTH, N. W. 1952. Suggested classification of muskeg for the engineer. Eng. J., Vol. 35, No. 11, pp. 1199–1210.

——— 1955. Organic terrain organization from the air (Altitudes less than 1000 ft.): Handbook No. 1, Defence Research Board, Dept. Nat. Defence, DR. No. 95, Ottawa, 49 pp.

——— 1962a. Organic terrain and geomorphology. Can. Geog., Vol. VI, No. 3–4, pp. 166–171.

———— 1962b. Why and how does muskeg occur. Proc. Eighth Muskeg Res. Conf., NRC, ACSSM Tech. Memo. 74, pp. 145–151.

THOMSON, S. 1957. Some aspects of muskeg as it affects the Northwest Highway System. Proc. Third Muskeg Res. Conf., NRC, ACSSM Tech. Memo. 47, pp. 42–49.

US GOVT. PRINTING OFFICE. 1961. Transport requirements for the growth of northwest North America. House Doc. No. 176, Vol. 2, pp. viii–2–36, ix–2–7, Washington, DC Res. Rept. Battelle Memorial Inst. on an integrated transport system to encourage economic development of Northwest North America.

WALSH, J. P. 1957. Some economic aspects of construction in muskeg areas, with particular reference to oilfield roads. Proc. Third Muskeg Res. Conf., NRC, ACSSM Tech. Memo. 47, pp. 33–41.

2 Classification of Muskeg

Chapter Co-ordinator
N. W. RADFORTH

2.1 NATURAL RELATION: THE BASIS OF CLASSIFICATION

In the methodology of reference to the phenomena, character, and dynamics of muskeg, engineering classification must satisfy requirements in two areas of interest: the muskeg condition itself and its two kinds of constituent material, peat and the overlying living vegetation. In both areas the natural relation (basically biotic), if used as the expression of organization, makes possible the prediction of terrain states and an intelligent design for engineering application. Utilization of an artificial (unnatural) approach to classification does not render it invalid but does invoke extensive explanation and interpretation because of the need for reconciliation with natural trends. Otherwise, prediction techniques would be cumbersome and oblique and have less than desirable reliability. Furthermore, a natural system of classification facilitates comparison and integration with other subsidiary or different approaches to classification which are used in other disciplines and are not emphasized in this Handbook. An example of the latter is the differentiation of peat by reference to pH or chemical content (Woods 1950).

In acknowledging muskeg condition as distinct from its constituent material, it is appreciated that gross differences in muskeg are often the expressions of integrated characteristics or the result of environmental character combined with that of the constituents. To justify classification of muskeg states, therefore, one must rely upon the classification of type features exemplifying peat and vegetal cover types and account somehow for the significance of the environments in which they exist. Classification, as presented here, conforms to this logic and presents first the hidden features, i.e. those characterizing the structure of the peat.

2.2 CLASSIFICATION AND PEAT STRUCTURE

Only two structural entities are appreciated in gross samples of peat. One is fibrous, the other is granular. These designations imply shape only and take no account of size, orientation, texture, or chemical constitution. They symbolize a value that is not necessarily botanical or even organic. It is important to understand that the very small granules may behave as colloidal particles, which implies that peat is partly colloidal. Varying degrees of the colloidal state must be expected. The water and dissolved compounds associated with peat are important physical constituents, but they are no more intrinsic than water would be in establishing the meaning of "clay."

In engineering application, some if not all qualifications to the descriptions of the constituents must be incorporated if either a qualitative or quantitative concept of peat structure is to be gained. For classification purposes it can be assumed that peat is entirely organic, even though wind- or water-conveyed mineral matter occasionally adulterates it. The texture of the elements has obvious significance, and an attempt has been made to explain this quality, particularly with reference to fibres, by MacFarlane and Radforth (1968), who proposed the name "axon" not only to convey linearity but also to provide a model upon which secondary physical attributes of the fibres could be imposed. Where the concentration of "axons" varies in characteristic ways in the peat sample, and distribution of cellulose (non-woody) and lignocellulose (woody) is involved, texture becomes significant as a structural quality. Laboratory analysis is necessary to assess texture if quantitative answers are required, but woodiness and non-woodiness in the gross sense can be determined qualitatively from hand samples in the field. A qualitative analysis is often all that is required for initial engineering reference and purpose.

For comprehensive interpretation, botanical (anatomical) knowledge of the two basic ingredients of peat is a help and encourages precision (Eydt 1962), together with an understanding of the environment from which the sample was taken and an appreciation of the range of textural difference occurring at the site. There are other advantages to the application of botanical knowledge, but these are referred to elsewhere (Radforth 1967).

2.3   FLORISTIC CONSTITUTION OF PEAT

Apart from texture, botanical inference occasionally has to be considered by the engineer if the only information available about the peat has been provided in botanical terms. Sometimes peat type is named in floristic terms. One of these is sphagnum peat, which designates peat containing 60 per cent or more of sphagnum moss (ASTM 1967). Sedge peat, another floristic designation, implies that the major botanical constituent is composed of one or more species of sedge (plants that are grasslike in appearance). Sphagnum peat has great water holding capacity and high insulating value in contrast with sedge peat, which has better binding qualities and in gross samples is less elastic. Neither constituent is woody. In the United States, when the fibrous condition is implied, the components involved may be either sphagnaceous or sedge-grassy, or both, in structural character, but in Britain sphagnaceous peat is designed separately and fibrous peat is always sedge peat. The confusion thus created can be largely avoided if physical values, no matter how qualitative, can be used instead of botanical ones. If it is necessary to use floristic names for general purposes – say, to effect broad understanding of the material – this has some engineering merit and explains the origin of the structural units. Physical value enables understanding and eventual computation in deriving mechanical properties. Texture in the sense claimed has significance for the evaluation of bearing strength, compaction and consolidation, elasticity, porosity, permeability, cohesion, and tensile strength. To these features, botanical derivatives contribute proportionately, but all of them are functions of physical quality.

2.4   STRUCTURE AND PROPERTIES OF PEAT

Both fibres and granules have variable and partial influence on the gross mechanical potential, depending upon the three-dimensional size of the elements. Fibres a foot (30.5 cm) or more in length are stronger than shorter fibres. This has great significance in the field. Under either static or dynamic load, long-fibre peat endures and sustains load under critical circumstances. Granules will respond physically in one way when colloidal in size, in another way when larger and participating as particles in a mechanical mixture. Both orders of size occur, often in the same sample, but in differing proportions. It is recorded that the largest linear elements constituting the primary matrix of peat are of tree-trunk proportions (MacFarlane and Radforth 1968). These elements, when less numerous and sporadic, are designated in the literature as "woody erratics," an expression not originally intended to signify any particular orientation of the element in the matrix. If the connotation is now allowed to apply for all values of frequency within the matrix, the term has wider usefulness.

There does not appear to be any reported method of assessing what provides the greatest contribution to strength and other mechanical properties of peat. Does this contribution come from the integrated mechanical effect of all the elements, or from the collective effect provided by special orientation and mechanical interrelation of both fibres and granules in different densities of mesh? To provide the best information on classification, information on gross arrangement should be given only after inspection of field samples is followed by structural analysis in the laboratory. If precise detail is required with statistical accuracy, no fewer than thirty samples from randomly selected sites in the muskeg involved should be examined. This procedure will assist in establishing data characterizing the predominating and characteristic structural arrangement with 80 per cent reliability (Ashdown and Radforth 1961). At the outset, the investigator is advised to determine the peat category (Radforth 1952) to which the samples belong. This step in classification will place the sample in the main peat type without recourse to botanical interpretation.

2.5   TYPIFICATION OF PEAT STRUCTURE

In order to specify and account for basic structural identity, it is necessary to provide nomenclature that is referable to the botanical source of the elements in the peat and to the biological control system governing its structure. Acceptance of this principle helps in describing and delineating homogeneity (or perhaps heterogeneity) in the peat, consistency and persistence of character, and recurrence of distributional phenomena. More important, adherence to this principle facilitates relation between peat and cover. The cover, obviously biologically governed and the "progeny" of the ancestral constitution characterizing the peat, accounts indirectly for the peat structure subtending it. If there is provision for cross-reference, and proper use is made of it, interpretative reference to cover will indirectly imply the required structural information about the peat. When it is appreciated that cover may be viewed readily from either surface or air whereas peat structure cannot, the advantages of the correlative terminology are obvious. This provision is a particularly useful corollary of the principle proposed.

Usually the typification required is for the peat deposit as a whole. Frequently this has to be attempted from a single core sample by an analyst who has not seen the deposit. In this case, the results of structural (macrofossil) analysis of the core will apply for the site from which the core was removed; but experience shows that extrapolation as to type beyond about 20 feet (6.10 m) from the core hole is dependent upon supportive evidence. Either more cores should be procured (which is sometimes impractical and expensive) or the existing sample should be microanalysed for additional evidence. Frequency relations of the fossil spores or cuticles procured from successive depths in the core can be taken as typical for a more extensive region than that defined by the structural (macrofossil) analysis and a region more extensive than that prescribed by the single core structural analysis (Radforth 1952).

In addition, macrofossil remains cannot always be identified botanically to provide the information needed to suggest whether the structural entity itself would have characteristically widespread distributional significance. Microfossil constitution gives less emphasis to highly local features atypical of the order of structural homogeneity than the observer sees in the main mass of peat for which microfossil characterization obtains. The microfossil conspectus thus offers an acceptable general basis for classification. It suggests (and also confirms) the composite macro-type. Moreover, the micro-type, through comparative study, facilitates comprehension of the macro-type in terms of biotic organization because identification is now in generic as well as structural terms, the generic evidence having been supplied by the microfossil factor.

When structural typification of the peat deposit is confirmed and complete, categorization of the types encountered is possible. Descriptions of the peat categories (Table 2.1) combine information about structural essentials in the peat

TABLE 2.1
Classification of peat structure

Predominant characteristic	Category	Name
Amorphous-granular	1	Amorphous-granular peat
	2	Non-woody, fine-fibrous peat
	3	Amorphous-granular peat containing non-woody fine fibres
	4	Amorphous-granular peat containing woody fine fibres
	5	Peat, predominantly amorphous-granular, containing non-woody fine fibres, held in a woody, fine-fibrous framework
	6	Peat, predominantly amorphous-granular containing woody fine fibres, held in a woody, coarse-fibrous framework
	7	Alternate layering of non-woody, fine-fibrous peat and amorphous-granular peat containing non-woody fine fibres
Fine-fibrous	8	Non-woody, fine-fibrous peat containing a mound of coarse fibres
	9	Woody, fine-fibrous peat held in a woody, coarse-fibrous framework
	10	Woody particles held in non-woody, fine-fibrous peat
	11	Woody and non-woody particles held in fine-fibrous peat
Coarse-fibrous	12	Woody, coarse-fibrous peat
	13	Coarse fibres criss-crossing fine-fibrous peat
	14	Non-woody and woody fine-fibrous peat held in a coarse-fibrous framework
	15	Woody mesh of fibres and particles enclosing amorphous-granular peat containing fine fibres
	16	Woody, coarse-fibrous peat containing scattered woody chunks
	17	Mesh of closely applied logs and roots enclosing woody coarse-fibrous peat with woody chunks

matrix and are sufficiently comprehensive and detailed to characterize the basis for identification of the recurring peat types.

Each category of the list of 17 is pictorially represented in a comparative series (Figs. 2.1–2.16) except for Category 17 (Fig. 2.17). In this category the woody erratics are so large and so frequent that a photograph to the same scale as those for the other 16 categories would show merely a surface view of part of a single piece of fossilized wood. A view of the peat category as it appears in the landscape has been provided instead. This photograph shows how large and numerous the woody erratics are. Although for the inexperienced observer the photographs are aids to identification, the help they afford is limited and they are least useful when isolated. When classifying, the observer must use all the photographs comparatively along with the associated descriptions. A decision on identification should be made only following a fully comparative inspection of all sources of information.

The choice of the expression "category" is deliberate. It is intended to imply that a range in proportion of the constituent materials is to be expected. Despite

FIGURE 2.1   Peat category 1

FIGURE 2.2   Peat category 2

FIGURE 2.3   Peat category 3

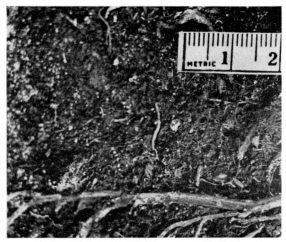

FIGURE 2.4   Peat category 4

FIGURE 2.5   Peat category 5

FIGURE 2.6   Peat category 6

FIGURE 2.7   Peat category 7

FIGURE 2.8   Peat category 8

FIGURE 2.9   Peat category 9

FIGURE 2.10   Peat category 10

FIGURE 2.11   Peat category 11

FIGURE 2.12   Peat category 12

FIGURE 2.13   Peat category 13

FIGURE 2.14   Peat category 14

FIGURE 2.15   Peat category 15

FIGURE 2.16   Peat category 16

this range, each category is distinctive and therefore acceptable. Every known "type" of peat fits naturally into one or another of these categories.

Category descriptions (Table 2.1) not only convey basic form–texture information, they suggest as well, but only in general terms, the primary feature of arrangement of the peat constituents. The precise orientation of fibres and their size relations are not represented in the descriptions. The opacity of the samples makes it difficult to describe the orientation without more than normal field inspection of the samples. Laboratory dissection involving mechanical sorting reveals the rough proportions of described constituents, but the sorting interferes with the natural orientation and arrangement patterns of the constituents.

FIGURE 2.17   Peat category 17

As a further aid to structural interpretation, Rockley and Radforth (1967) attempted to apply X-ray photography of samples taken with a Davis sampler. The photograph of the peat type for category 6 (Fig. 2.6) shows confusion between non-woody, woody fine-fibrous, and amorphous-granular constituents. An X-ray photograph shows woody coarse fibres to be distinct and tortuous, travelling in no particular plane (Fig. 2.18). Non-woody fine-fibrous elements appear to lie usually in a horizontal plane and to alternate as layers with shallow plates of amorphous-granular.

To symbolize the constituents graphically, diagrammatic configurations are shown in Figure 2.19. Combinations of these symbols in appropriate proportions make it possible to convey proportions of constituents and to differentiate pictorially between peat types (cf. dial configurations in Fig. 2.19(a), (b)). A peat category may include several types, but a type cannot belong to more than one category. For the present it is customary to accept a peat type as a commonly recurring subcategory. Until or unless there are types in certain or all categories that have special significance in either the scheme of biological order or in engineering application, no systematic value is assigned to them.

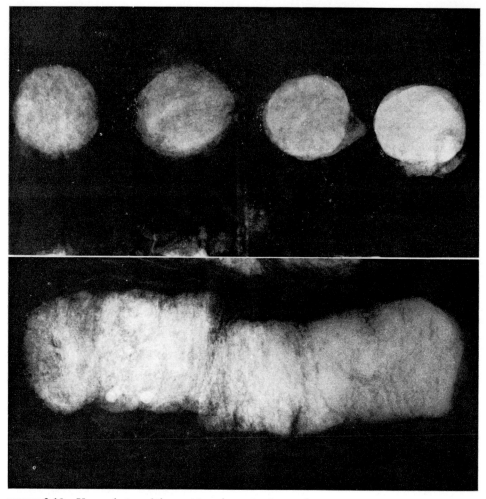

FIGURE 2.18   X-ray photos of the peat type in peat category 6

2.6 STRUCTURAL TYPIFICATION OF MUSKEG COVER

The method of reference used here to describe cover is referred to by botanists as based on life form. Usually botanical systems are designed to describe or account for plant communities and are not directly applicable to engineering use. Dansereau (1951) and Dansereau and Segadas-Vienna (1952) developed this kind of phyto-sociological system based on life form. If in engineering application – say, to drainage – there is need for intensive study on distribution of plant communities in muskeg, Dansereau's system or its equivalent is recommended. The system recommended here, on the other hand, is direct and has the advantage of revealing

TABLE 2.2
Properties designating nine pure coverage classes

Coverage type (class)	Woodiness vs. non-woodiness	Stature (approximate height)	Texture (where required)	Growth habit
A	Woody	15 ft or over	—	Tree form
B	Woody	5–15 ft	—	Young or dwarfed tree or bush
C	Non-woody	2–5 ft	—	Tall, grasslike
D	Woody	2–5 ft	—	Tall shrub or very dwarfed tree
E	Woody	Up to 2 ft	—	Low shrub
F	Non-woody	Up to 2 ft	—	Mats, clumps, or patches, sometimes touching
G	Non-woody	Up to 2 ft	—	Singly or loose association
H	Non-woody	Up to 4 in.	Leathery to crisp	Mostly continuous mats
I	Non-woody	Up to 4 in.	Soft or velvety	Often continuous mats, sometimes in hummocks

evaluations which relate the cover to the subsurface structural conditions already described.

Detailed examination of the constituents of the peat types represented in Table 2.2, Figure 2.19(a), (b), and comparison of them with the cover suggests that a given peat associates with characteristic cover composition, the classification of which will now be explained.

The nine classes of vegetal cover listed in Table 2.2 and referred to in column 1 by the letters A to I are, like the peat categories, symbols of and functions of the structure in that they refer to the shape and texture of the individual elements making up the composition. In addition, they are indicative of size in that they refer to the height of the individuals. They also have reference to habit, the equivalent of arrangement. This feature was implied in the description of peat categories. Thus each class is a characteristic entity conceived on structural evidence and in which limited range of variation occurs in one or several of the structural components.

The occurrence of pure classes in nature is rare; association of classes is more normal. To indicate proportionate representation of class, the letter symbols are combined into formulae. A formula is constructed as the observer in the field views the area which he has already delineated geographically; he mentally circumscribes the area. The predominating class of cover is usually, but not necessarily, the tallest. The components of lesser stature appear in the formula after the one indicating predominance, in the order of the relative lesser amounts of cover they constitute. If the estimated percentage of cover for any class is less than 25 per cent, it should not appear in the formula. Therefore, since four classes of cover are

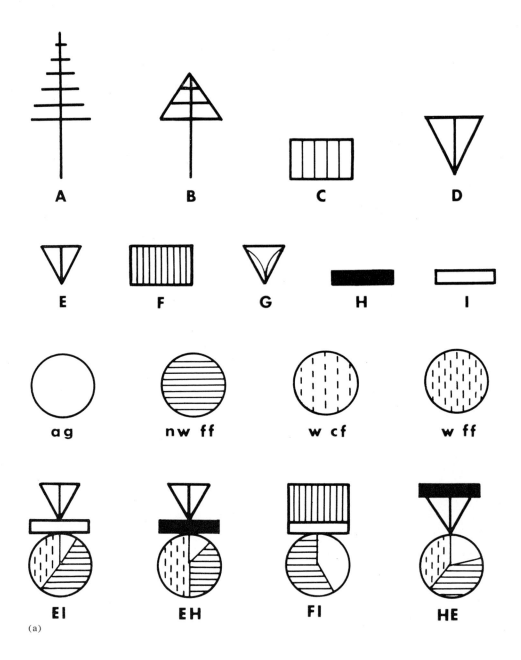

A   B   C   D

E   F   G   H   I

a g   nw ff   w cf   w ff

EI   EH   FI   HE

(a)

FIGURE 2.19(a), (b)   The commonly recurring cover formulae and associated peat types presented diagrammatically (cf. Table 2.2). Subsurface constitution: w cf, woody coarse-fibrous; w ff, woody fine-fibrous; nw ff, non-woody fine-fibrous; ag, amorphous-granular

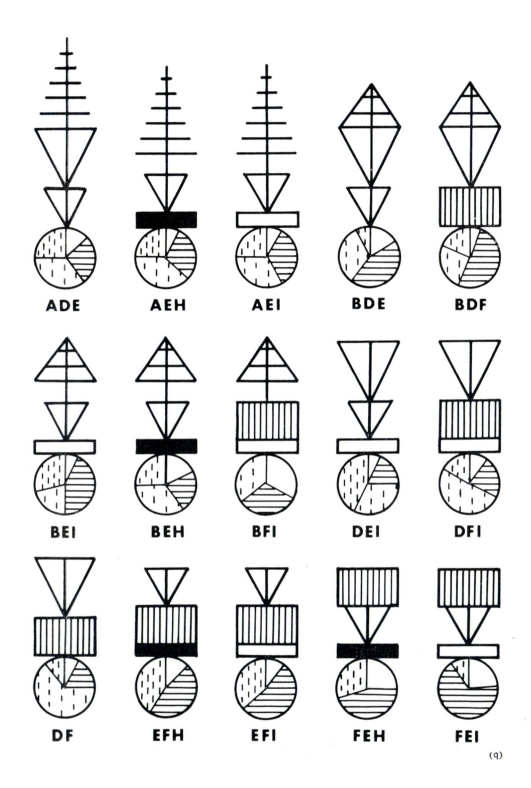

ADE AEH AEI BDE BDF

BEI BEH BFI DEI DFI

DF EFH EFI FEH FEI

(9)

statistically unlikely to appear in exactly equal proportions of 25 per cent, no formula will show more than three classes. In many instances only two classes comprise a formula.

Within the range of 25 per cent no attempt is made to provide secondary levels of class, because those classes with a share greater than 25 per cent constitute together the characterizing contribution of the structure of the underlying peat. It is known that this characterizing contribution, when completely fossilized, does not correspond exactly with cover class percentages. Some components of cover structure fossilize more readily and contribute more bulk to the peat than proportions of class factors in the cover formulae suggest. For a given cover formula, however, the end product of structural contribution is consistently the same.

When classifying in the field, one is apt to forget that class symbols represent composite ideas. Stature is very important in the operation of classifying, but form, habit, and texture are equally important though perhaps less obvious to the observer. All these qualities share prominence with proportionate amounts for each cover class. A proportionate amount of class representation in cover formulae is as important as the fact of class presence in the formula.

2.7   INTERPRETATIVE VALUE OF SYMBOL SYSTEM

Identification, achieved in the performance of selecting the appropriate symbols to characterize site, is an end in itself. Engineering use of this information affords further gains if interpretative implication is exploited. Each cover class indirectly signifies the occurrence of prevailing conditions which may be taken into account for purposes of engineering design. These conditions are listed in Table 2.3.

Certain formulae signify qualification for the claims indicated in the several conditions of Table 2.3: (a) where Class A is prominent but followed by F in the formula, the peat is less well drained or drainable than if Class E follows; (b) where Class A is involved in a formula with H, drainage conditions are better throughout the year than for conditions in item 3 under Class A in Table 2.3; (c) the situation expressed in items (a) and (b) also obtain when, in those circumstances, B replaces A.

2.8   GRAPHIC SYMBOLS FOR TOTAL COVER TYPIFICATION

Understanding of the meaning conveyed in formulae comprised of the class symbols can be enhanced if diagrammatic synopses accompany the formula. Table 2.2 and Figures 2.19(a), (b) show proposed graphic characterization of the letter symbols. They also indicate how combinations of diagrams appear in association with their corresponding cover formulae. The combinations selected represent the commonest structural configurations and are shown in association with the corresponding diagrammatic expressions depicting the peat structure types.

TABLE 2.3
Occurrence of prevailing conditions for the various cover classes

Predominant class in formula	Engineering significance
A	(1) Presence of large woody erratics in the peat (2) The position of relatively shallow depths of peat for the landscape as a whole (3) Location of best drained peat (qualified by items (a) and (b) in text) (4) Location of best drained mineral soil sublayer (5) Presence of highly permeable peat (6) Vicinity of lowest summer temperatures in the peat (7) Location of the coarsest, most durable peat (8) Best conditions for static load (see, however, item (c) in text) and dynamic loading
B	Same as for A above, but less intensively represented
C	Predominance rare, except in tropical and subtropical locations (for example, Guyana, Brazil, Paraguay and Uruguay, and possibly Southern Rhodesia, Nigeria, Israel, Malaysia, etc.)
D	(1) Linear drainage, often an open water course (2) Lagg condition around a confined muskeg (bog) (3) Traps present (4) Good, but highly elastic, bearing conditions; difficult to consolidate and with marked patterned local differentials as to rate of consolidation (5) Features highly conducive to spring flooding (6) Silt in the mineral soil sublayers with highly mixed aggregate from outwash (7) Features conducive to differential settlement (often abrupt) under load
E	Equally important as D above and is very common in temperate, arctic, and subarctic zones (1) High order of homogeneity in peat, even in relation to micro-topography in which mounds, ridges, and ice knolls are important (2) Peat difficult to re-wet once drained of gravitational water (3) Conditions accommodating to certain articulated wheeled vehicles (Enright 1962) (4) Good cohesion and tensility, moderate elasticity even when water shows at the surface in the field (5) Easily drainable conditions (for free water)
F	Presence of highly critical conditions when prominent in the formula: (1) Low points on drainage gradients (2) Muskeg with centres of extremely low bearing potential whether wet or relatively dry (3) Peat of low tensile strength and showing little elasticity unless the local water table is consistently high (small open pools the year round) (4) Sites where shear strength is lowest in muskeg at frequent intervals; water is not excessive
G	Rarely predominates in the formula, is indicative of a highly fluctuating water table
H	When predominant indicates presence of: (1) Permafrost and late seasonal subsurface ice conditions of uneven contour (2) Maximum range of microtopographic amplitude (often abrupt) for all muskeg (3) Local imponding and highly irregular, dissected drainage gradients (4) Relatively locally degraded peat (structurally and mechanically disrupted (Eydt 1962))
I	Unless Class I is the only component comprising the cover formula (which is rare), it lacks prominence. When it is a single contributing factor in cover it is very local, usually no more than 4 or 5 m in area of coverage, and the following occur: (1) Vehicle immobilization on the second pass for amphibious vehicles (2) The base of minor or major drainage gradients

2.9 COMPLEMENTARY SYSTEMS OF MUSKEG REFERENCE

The British Mires Research Group, realizing that characterization of structural features in vegetal cover of peatland provides the lead to identification of terrain characterization in the sense appreciated in the present approach, recently developed another system. Their system, still tentative, is referred to in Table 2.4. A combination of this system with that shown in Table 2.2 has been devised and presented in Table 2.5. The latter combines the essential values from both systems without being overcomplex. It is suggested that, even though the compromise is not used, it at least affords the means whereby the two fundamental systems (which arose independently) can be correlated.

The combined system allows for the use of the class symbols now widely adopted in Canada (cf. Keeling 1961; Enright 1962) which are known to obtain for other areas of the world where peatlands occur (Radforth 1967). There is some advantage in using symbols because, if they are available, experience shows that field observers quickly learn how to use the letters without recourse to the details of description conveyed by them. This promotes rapid and convenient forms of note-taking and recording in comparative study and in communication among observers, field operators, and laboratory personnel. It also affords a medium of convenience in the comparison of conditions observed by different workers in widely separated areas and is useful in techniques of mapping or producing overlays for airphoto mosaics.

2.10 COVER DESIGNATION AND TERRAIN PROPERTIES

Microfossil constitution provides the ultimate key by which relations between vegetal cover and macro-structure in the peat are established. It also provides the

TABLE 2.4
Vegetation (structure)

	Percentage of each structural type*					
	0–10	10–30	30–50	50–70	70–90	90–100
Trees over 5 m						
Trees under 5 m						
Shrub habit 50 cm–2 m						
Shrub habit up to 50 cm						
Creeping shrub up to 50 cm						
Broad-leaved herbs						
Sedge-graminoid habit 1–3 m						
(a) mats						
(b) hummocks						
Sedge-graminoid habit under 1 m						
(a) mats						
(b) hummocks						
Moss habit						
Lichen habit						

*Observer to insert percentages.

TABLE 2.5
Structural classification of vegetal cover of muskeg (peatland)
(integration of British and Canadian systems)

British Mires Research Group*	Class symbol	Texture	Radforth system		
			Stature	Form	
Trees over 5 m	A	Woody	15 ft or over (4.5 m or over)	Tree form	
Trees under 5 m	B	Woody	5–15 ft (1.5–4.5 m)	Young or dwarfed tree or bush	
Shrub habit 50 cm to 2 m	D	Woody	2–5 ft (0.6–1.5 m)	Tall shrub or very dwarfed tree	
Shrub habit up to 50 cm Creeping shrub up to 50 cm	E	Woody	Up to 2 ft (up to 0.6 m)	Low shrub	
Broad-leaved herbs	G	Non-woody	Up to 2 ft (up to 0.6 m)	Singly or loose association	
Sedge-graminoid habit 1–3 m (a) mats (b) hummocks	C	Non-woody	2–5 ft (0.6–1.5 m)	Tall, grasslike	
Sedge-graminoid habit under 1 m (a) mats (b) hummocks	F	Non-woody	Up to 2 ft (up to 0.6 m)	Mats, clumps, or patches sometimes touching	
Moss habit	I	Non-woody	Up to 4 in. (up to 10 cm)	Often continuous mats, sometimes in hummocks	
Lichen habit	H	Non-woody	Up to 4 in. (up to 10 cm)	Mostly continuous mats	

NOTE: Following classification, observer states percentage of cover class within 20%.
*Adapted by Radforth.

link with natural attributes. Although classification of subsurface mechanical composition made indirectly by reference to structural composition in vegetal cover is on a non-floristic basis, explanations of relation – if required – are floristically entailed through the medium of microfossils.

It is possible in very general terms to place additional quasi-quantitative values on qualitative structural composition as developed to this point. MacFarlane (1968) attempted to assess peat type with the use of a shear vane, but identity of type for classification purposes is difficult to establish because shear values are functions not only of structure but also of water content. Furthermore, the size of vanes relative to the arrangement of the structural elements in the peat is difficult to optimize and the values obtained are highly empirical (see Section 5.2(1)). On the other hand, the results suggest the effect of water and mechanical degradation of peat type in relation to depth for peat beneath specified cover formulae. The results also differentiate peat types in depth and in association with accompanying water but on a relative basis only.

Other attempts at demonstrating the strength of the peat-cover complex were made through the application of a field instrument which became known as a "fluke" or "anchor." This instrument (Fig. 2.20) is secured to the muskeg mat by

FIGURE 2.20   A muskeg fluke inserted in the cover and underlying peat to determine strength values when the terrain complex undergoes shear

its steel rods, and the mat will shear when an appropriate force is applied to an attached cable. If a dynamometer is attached to the cable, tension values can be read off at the time the mat shears. Table 2.6 shows comparative values obtained for different cover formulae. Their usefulness becomes apparent for purposes of identification only when it is appreciated that cover formulae indicate the presence of range in strength value for a given peat category. Individual values may differ with differences in water content.

TABLE 2.6
Fluke test results on muskeg (August 1961, Parry Sound, Ont.)

Test site			Average shearing
Area	VTS*	Cover formula	force (lb)
1	1	FI, wet between EI mounds	2100
1	1	EI mounds, E ∼ I	1650
1	2	FIE	2450
1	3	EI mounds, E ∼ I	1467
1	3	FI (low, wet area)	2667
2	1	FI (very wet), dense F	2483
2	1	EI mounds, dense E	2788
2	2	IF, I very dense	1700
2	2	EI mounds, E = I	1933
5	1	DFI (very wet)	1950
3	1	FI (very wet)	1650
3	1	EI mounds, E = I	2050
4	2	FI, F = I	2417
4	2	IE hummocks	1717
4	3	FIE, F = I	2367
4	3	EI hummocks	2600

*VTS = Vehicle test site.

Similar attempts at establishing identity on a quantitative basis have involved the use of the cone penetrometer (cf. Rush, Schreiner, and Radforth 1965). Without some means of correcting for the influence of water and possible secondary compression phenomena, it is difficult to demonstrate characterization in accordance with cover or peat category.

Strength (shear) values regularly increase with depth, but the degree to which this change occurs is now known to depend upon the local prevalence of gravitational water in the terrain. In well-drained muskeg, cone penetrometer values generally double with depth for given cover formulae. This relation is hard to demonstrate however for cover formulae; the values are erratic and show great deviation from the average (OATRU, McMaster University 1966).

If the water factor is considered on the basis of macro-climate rather than micro-climate, for a given cover formula there are wide differences in the strength of peat type as expressed by cone penetrometer values. From comparisons of values taken at 3-inch (7.62-cm) intervals of depth, peat beneath EI cover deposits are considerably stronger near the surface in the Falkland Islands (unconfined muskeg) than they are in the Parry South district of Ontario (confined muskeg). The same

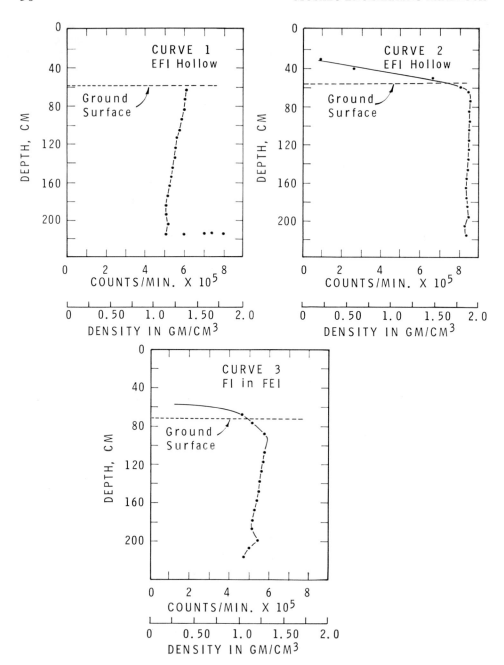

FIGURE 2.21   Peat density measurements, Parry Sound, Ontario

relation holds for peat deposits beneath FI cover in the two areas, but do not hold where FEI is encountered. For FEI, Parry Sound peats were stronger up to a depth of 2 feet (61.0 cm), after which strengths were comparable for the two areas. For peat below 2 feet, greater strength was experienced in the Parry Sound peat than in the Falkland Island peat. In these analyses the local conditions with respect to free water in the terrain were judged comparable on a qualitative basis. The peat per given structure category is denser and harder where water sources are largely from high humidity and precipitation rather than from lateral inflow of water in the ground.

Recently, a nuclear density-moisture probe for measuring relative differences in density of peat type was devised (Radforth and Radforth 1965) and is still being tested. This instrument shows great promise for assessing the influence of water on density. But even before the water factor is accounted for, total density in the mass can vary with the cover formula as expressed in Figure 2.21. Subsequent work has shown, however, that peat density under a given cover category has a range of values, and that the ranges for different cover categories overlap to some degree. To the extent that characteristic delineation is possible, density difference is a feature of identity and can be incorporated into the general classification system with other physical measurements derived on the quasi-quantitative basis.

The allusion to water and climate as bearing on classification suggests that, once the materials (peat and cover) are classified by a method incorporating reference to observable features in the field, the physiographic connotation in muskeg can be accounted for in classifiable terms. Thus, the geographic extent and shape of areas, characterized according to a given cover formula can be mapped. The shape of the areas so characterized indicates ground-water deployment and micro- and macro-configurations which relate to the natural aspects of order symbolized by the pertinent formulae (MacFarlane 1958).

There are still features of order that have not been accounted for. Deployment of subsurface ice (Radforth 1954, 1961), permafrost occurrence, geomorphic difference (Radforth 1968), and major drainage systems are the end reflections of organization involvement yet to be identified and classified in both confined and unconfined states. Some may argue that extension of the system so far into environmental values (extra biotic) requires application not only of classification but also of interpretation. This aspect is emphasized in chapter 3 and the broadest aspects of muskeg character will be dealt with there.

REFERENCES

AMERICAN SOCIETY FOR TESTING AND MATERIALS (ASTM). 1967. Rules on packaging peat and peat moss. Proposal submitted to Nat. Conf. on Weights and Measures.

ASHDOWN, K. H. and N. W. RADFORTH. 1961. The procurement of physical and mechanical data for organic terrain. Proc. Sixth Muskeg Res. Conf., NRC, ACSSM Tech. Memo. 67, pp. 88–104.

DANSEREAU, P. 1951. Description and recording of vegetation upon a structural basis. Ecology, Vol. 30, No. 2, pp. 172–229.

DANSEREAU, P. and F. SEGADAS-VIENNA. 1952. Ecological study of the peat bogs of eastern North America. Can. J. Botany, Vol. 30, pp. 490–520.

ENRIGHT, C. T. 1962. The muskeg factor in the location and construction of an Ontario Hydro service road in the Moose River basin. Proc. Eighth Muskeg Res. Conf., NRC, ACSSM Tech. Memo. 74, pp. 42–58.

EYDT, H. R. N. 1962. An assessment of the component tissues of peat in their *in situ* arrangement. Unpubl. PHD thesis. McMaster University, Hamilton. 89 leaves.

KEELING, L. 1961. The organic terrain factor and its interpretation. Proc. Seventh Muskeg Res. Conf., NRC, ACSSM Tech. Memo. 71, pp. 102–126.

MACFARLANE, I. C. 1958. Guide to a field description of muskeg (based on the Radforth Classification System). NRC, ACSSM Tech. Memo. 44, rev. ed., 36 pp.

——— 1968. Engineering studies of peat soils (muskeg). Trans., Second Internat. Peat Congress (Leningrad, 1963), Vol. II, pp. 853–862, H.M.S.O., Edinburgh.

MACFARLANE, I. C. and N. W. RADFORTH. 1966. Structure as a basis of peat classification. Unpublished manuscript.

ORGANIC AND ASSOCIATED TERRAIN RESEARCH UNIT (OATRU). 1966. Unpublished field notes. Muskeg Lab., McMaster University, Hamilton.

RADFORTH, N. W. 1952. Suggested classification of muskeg for the engineer. Eng. J., Vol. 35, No. 11, pp. 1199–1210.

——— 1954. Paleobotanical method in the prediction of subsurface summer ice conditions in northern organic terrain. Trans. Roy. Soc. Can., Vol. XLVIII, Series III, Section Five, pp. 51–64.

——— 1961. Organic terrain. *In* Soils in Canada, *edited by* R. F. Legget. University of Toronto Press. Roy. Soc. Can., Spec. Publ. 3, pp. 113–139.

——— 1967. Muskeg and climate. Paper presented at Can. Bot. Assoc. Meetings, Ottawa (manuscript).

——— 1968. The principles of airphoto interpretation as related to organic terrain. Trans., Second Internat. Peat Congress (Leningrad, 1963), Vol. I, pp. 231–240, H.M.S.O., Edinburgh.

RADFORTH, N. W. and J. R. RADFORTH. 1965. The significance of density as a physical property in peat deposits. J. Terramechanics, Vol. 2, No. 3, pp. 81–88.

ROCKLEY, J. C. and N. W. RADFORTH. 1967. Structural interpretation of peat by X-ray techniques. Unpublished manuscript.

RUSH, E. S., B. G. SCHREINER, and N. W. RADFORTH. 1965. Trafficability tests on confined organic terrain (muskeg): Rep. 2, volume 1: Summer 1962 tests. U.S. Army Corps of Engrs., Waterways Exp. Sta., Tech. Rept. No. 3–656, pp. 1–48. Vicksburg, Miss.

WOODS, A. B. 1950. Unpublished data from work-book at Muskeg Laboratory, McMaster University, Hamilton.

3 Airphoto Interpretation of Muskeg

Chapter Co-ordinator
N. W. RADFORTH

3.1 AIRFORM PATTERNS

Inspection of muskeg cover at ground level does not readily reveal the presence
or absence of organization beyond that suggested by cover formulae. On unconfined muskeg contiguous cover formulae show a random distribution in which
characterization is not apparent. It is not until the association of cover formulae
is viewed from the air that evidence of other kinds of organization can be appreciated.

The most convincing demonstration of how organization reveals itself and
provides the basis of large-scale patterned landscape comes from examination of
aerial photographs taken at successive levels of altitude (Figs. 3.1–3.4). Vehicle
tracks may be noted through FI muskeg in Figure 3.1. In the perspective represented by the 5000-foot (1525-m) altitude photograph (Fig. 3.2), they facilitate
orientation for the interpreter whose attention also turns to the development of a
patch pattern. This pattern resembles terrazzo and consequently it has been
given the designation "Terrazzoid" (Radforth 1956a, b). The full visual effect
of the pattern can be appreciated best at 17,000 feet (5185 m) (Fig. 3.4). Even in
the photograph taken at 10,000 feet (3050 m) (Fig. 3.3) the pattern attracts the
attention of the observer perhaps more than any other feature, and there is no
difficulty about its identification.

It is important to emphasize that this pattern recurs, and that in Canada it is
found almost without exception in the northern part of the Northwest Territories
and in northern Quebec. It is now known to occur in northern USSR near the Ob
River, and is commonest where permafrost is continuous on lands where the
underlying mineral terrain is almost flat, or where, if irregularities do occur, they
are infrequent and low in contour (usually 2 to 3 feet (0.61 to 0.91 m)).

Farther south, where Terrazzoid is lacking, if the hidden mineral terrain is
largely featureless, the cover is likewise largely featureless, i.e. textureless. In these
circumstances there is little outcropping of underlying mineral terrain and where
it does occur, the outcrop is low, with long, gently rising shoulders. The featureless-
ness is so marked, especially from altitudes of 10,000–30,000 feet (3050–

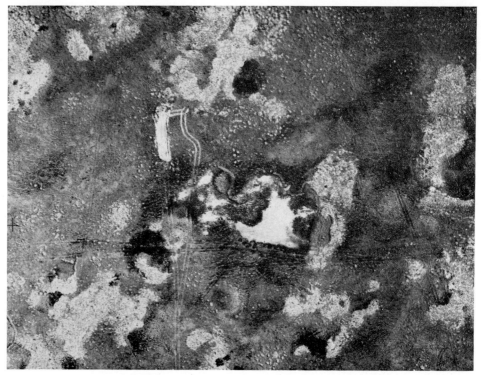

FIGURE 3.1   Terrazzoid airform pattern near Churchill area from 1000-ft altitude

FIGURE 3.2   Terrazzoid airform pattern near Churchill area from 5000-ft altitude

FIGURE 3.3   Terrazzoid airform pattern near Churchill area from 10,000-ft altitude

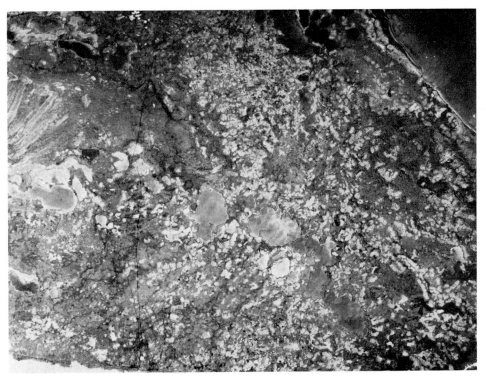

FIGURE 3.4   Terrazzoid airform pattern near Churchill area from 17,000-ft altitude

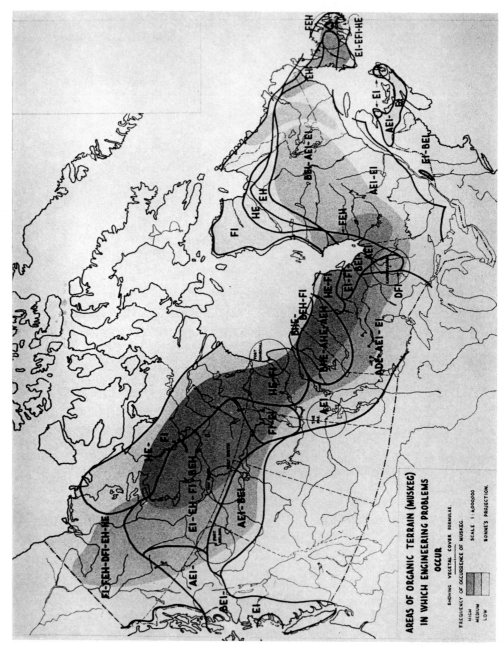

FIGURE 3.5 A map of Canada showing distribution of areas where common cover formulae occur most frequently (published with permission of the Defence Research Board, Canada Department of National Defence)

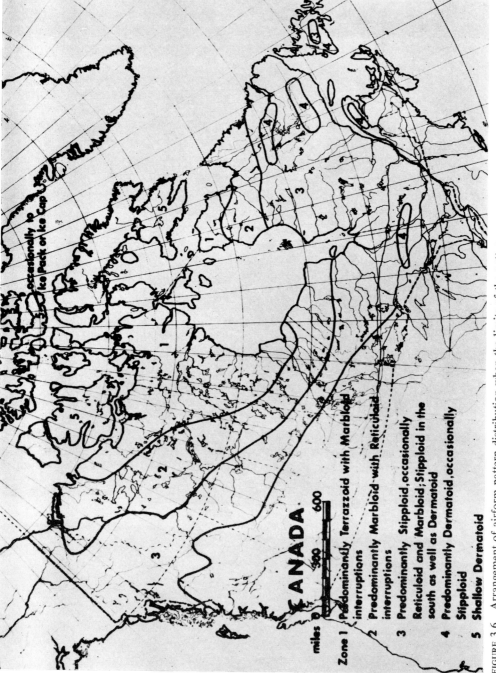

CANADA

miles 0   300   600

Zone 1   Predominantly Terrazzoid with Marbloid interruptions

2   Predominantly Marbloid with Reticuloid interruptions

3   Predominantly Stipploid, occasionally Reticuloid and Marbloid; Stipploid in the south as well as Dermatoid

4   Predominantly Dermatoid, occasionally Stipploid

5   Shallow Dermatoid

FIGURE 3.6   Arrangement of airform pattern distribution when the limits of the patterns are estimated in accordance with present knowledge

9150 m), that the condition has been named "Dermatoid" (skinlike) (Radforth 1956a, b). If water is relatively plentiful in this environment, the vegetal cover on the peat is FI; if less plentiful, its structure appears as EI. Dermatoid, unlike Terrazzoid, has maximum distribution in the south, usually south of the discontinuous permafrost zone.

Evidence of a third expression of organization is in the form of a marbled pattern. The components of pattern are highly convoluted and make frequent contact. This pattern was named "Marbloid." It is common in muskeg-covered terrain in which permafrost exists, but it may occur where permafrost is discontinuous. In this circumstance it is isolated and not easily discernible.

There is a fourth pattern, with a background usually featureless, superimposed with stippling in various arrangements, called "Stipploid." Although it does occur in permafrost country, it is commoner to the south of it.

Finally, a fifth pattern reveals itself as composed of a network or reticulum of terrain elements. The "net" is normally associated with open water which is distributed in characteristic fashion. This pattern is called "Reticuloid." It occurs independently of permafrost and is found as far north as Terrazzoid.

These patterns, when first described, were designated "Airform Patterns" (Radforth 1956a, b) (Figs. 3.7–3.14). They recall similar associations of components of certain other natural media. Consequently, inexperienced observers can easily recognize these patterns. On a map of Canada showing the distribution of areas where common cover formulae occur plentifully (Fig. 3.5), areas where each airform pattern occurs in highest frequency can easily be superimposed (Fig. 3.6).

In the designation of airform pattern, it is concluded from Handbooks Nos. 1 and 2 published earlier (Radforth 1955, 1958) that altitude is important. Handbook No. 1, describing experience gained through direct observation of the terrain for altitudes up to 5000 feet (1525 m), deals with a relatively narrow geographical perspective and the pattern is therefore in large scale. Elements of the pattern seen at 20,000–30,000 feet (6100–9150 m) (Radforth 1956b), in terms of indirect (airphoto) inspection, represent a scale of 1 mile to the inch (approx.).

3.2   LOW-ALTITUDE AIRFORM PATTERN

When Reticuloid is viewed at altitudes lower than 5000 feet (1525 m), secondary features of organization may become evident. The components of the reticulum are of three orders of length, the longest being greatly attenuated (Fig. 3.7). Decreasingly lesser degrees of attenuation will be noted in Figures 3.8 and 3.9 respectively.

To facilitate classification and reference, these three kinds of reticula, the components of which are tortuous, have been named Vermiculoid I, II, and III (wormlike) (Radforth 1958).

Certain other airform patterns, when viewed at altitudes of less than 5000 feet

(1525 m) reveal elements of organization not appreciated at high altitude. The stippled effect in Stipploid takes on a third dimension which can be interpreted in terms of stature and form. To emphasize this feature, the units involved are referred to as "apiculations" and the collective condition is designated "Apiculoid" (Fig. 3.10).

Marbloid when viewed at low altitude reflects considerable relief, and the marbled units appear to have depth and sometimes to overlap as do cumulus clouds. The detailed pattern was, therefore, named "Cumuloid" (Fig. 3.11).

Terrazzoid reflects its basic character when viewed from altitudes of less than 5000 feet (1525 m) and more than 1000 feet (305 m), but secondary attributes also appear. The lighter units in Terrazzoid are raised "islands" in an open rough-textured background which they appear to interrupt. The high frequency of interruption suggests the name "Intrusoid" (Fig. 3.12).

Dermatoid from altitudes between 5000 and 1000 feet remains textureless in that the vegetation is not broken either in form or (significantly) in stature. For this reason the condition was designated "Planoid" (Fig. 3.13), but one can detect changes in the colour of the vegetation across an area.

One feature of the secondary pattern which arises in permafrost country and as a relic beyond the zone of permafrost discontinuity is the closed polygon (Radforth 1958). Sometimes it overlaps Cumuloid, Terrazzoid, Dermatoid, or Apiculoid. "Polygoid" (Fig. 3.14) is the term used for this condition.

Definitions of these low-altitude airform patterns are given in Table 3.1, and mapping symbols are shown in Fig. 3.15.

3.3 PREDICTION

Airphoto interpretation leading merely to identification of terrain features (for example, airform patterns), though basic, is of little use to the engineer without additional analysis. The next step involves interpretation directed towards user requirement, with the primary airform pattern as the medium on which interpretation

TABLE 3.1
Description of airform patterns for 1000–5000-foot altitudes over organic terrain

Planoid	An expanse lacking textural features; plane
Apiculoid	Fine-textured expanse; bearing projections
Vermiculoid	Striated, mostly coarse textured expanse; featured markings tortuous
Vermiculoid I	Striations webbed into a close net and usually joined
Vermiculoid II	Striations in close association, often foreshortened and rarely completely joined
Vermiculoid III	Striations webbed into an open net, usually joined and very tortuous
Cumuloid	Coarse-textured expanse with lobed or fingerlike "islands" prominent; components shaped like cumulus clouds
Polygoid	Coarse-textured expanse cut by intersecting lines; bearing polygons
Intrusoid	Coarse-textured expanse caused by frequent interruptions of unrelated, widely separated mostly angular "islands"; interrupted

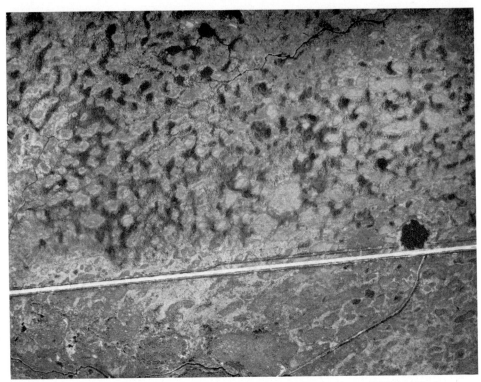

FIGURE 3.7   Vermiculoid I airform pattern near the Churchill area from 5000-ft altitude

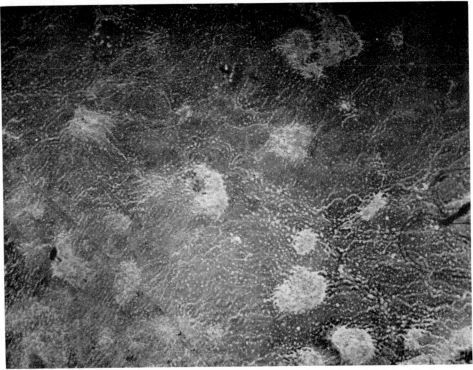

FIGURE 3.8   Vermiculoid II airform pattern near the Churchill area from 1000-ft altitude

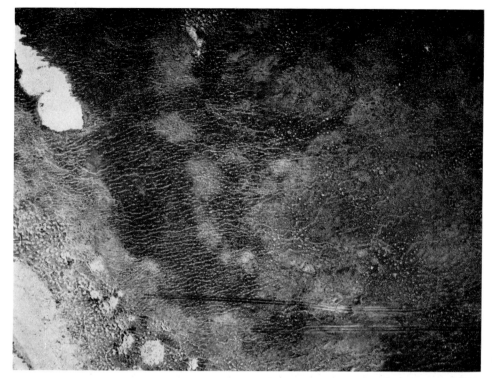

FIGURE 3.9   Vermiculoid III airform pattern near the Churchill area from 1000-ft altitude

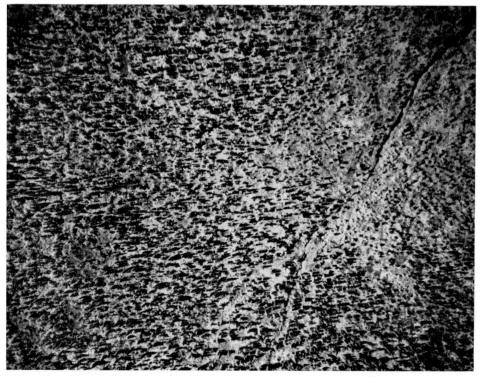

FIGURE 3.10   Apiculoid airform pattern near the Churchill area from 1000-ft altitude

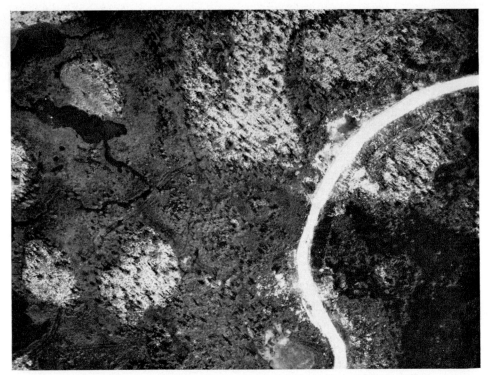

FIGURE 3.11   Cumuloid airform pattern near the Churchill area from 1000-ft altitude

FIGURE 3.12   Intrusoid airform pattern near the Churchill area from 1000-ft altitude

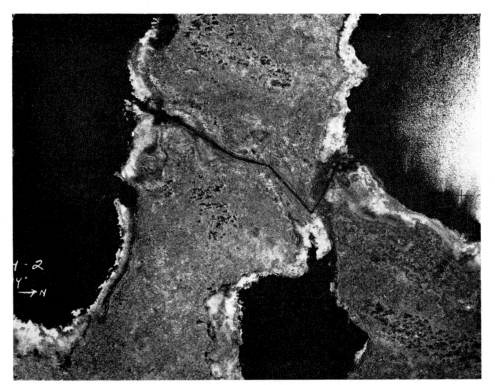

FIGURE 3.13   Planoid airform pattern near the Churchill area from 1000-ft altitude

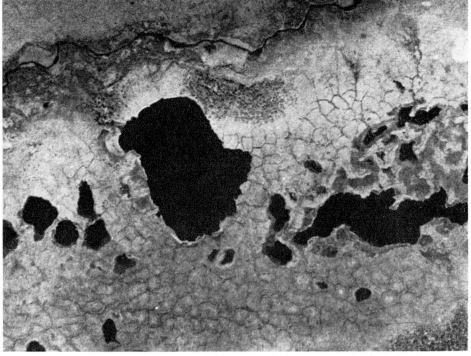

FIGURE 3.14   Polygoid airform pattern near the Churchill area from 1000-ft altitude

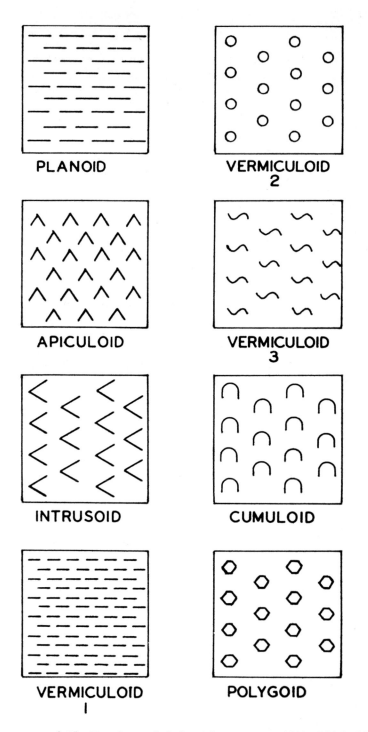

FIGURE 3.15 Mapping symbols for airform patterns, 1000–5000-ft altitudes (after the author)

should be made. To anticipate every engineering need is virtually impossible, but it will help to propose a number of broad semi-hypothetical examples, each general in implication yet narrow enough to be of practical use. It is intended that each example will offer direction for a wider spectrum of common but subsidiary or related types of engineering need.

In the first of these examples, the site is in Precambrian Shield country where granitic folds have precipitous sides. The centre of reference is the cover formula FI classified either from direct low-altitude inspection (under 5000 feet (1525 m)) or the equivalent in an airphotograph by application of interpretative knowledge. If the latter approach is used, determination of the cover begins by recognizing the presence and distribution of medium gray texture in the photograph (cf. Radforth 1955). The airform pattern (Dermatoid) will show distribution of colour (patchiness) in irregular "blotchy" admixtures with slightly darker gray background, which represents relatively drier muskeg with another kind of cover (transient and casual in influence).

Perhaps the predominating cover formula (FI) is already known from a site examination. If airform pattern is required to effect wider interpretation, the use of collateral evidence derived from ecological and physiographic features will lead to the designation of Dermatoid. For Stipploid, Terrazzoid, or Reticuloid, experience has also confirmed that the structure of these patterns as it appears on airphotos would be correspondingly characteristic and recognizable from the ground. If inspection is first made from the ground, the beginner will have some difficulty deciding on the pattern because of either inadequate perspective or overdevotion to, and misplaced emphasis on, local detail.

The combination of the relatively wet condition of the FI with the slope, disposition, and kind of associated mineral terrain limiting the Dermatoid pattern enables the observer to conclude that the water promoting peat formation is from run-off and/or springs, probably from the latter. The hydrology is defined by the expression "horizontal fen." Fen implies that most of the flood or retained water originates from the surrounding mineral terrain.

If the disposition and frequency of the surrounding upland elevations compare favourably with the granitic folds of the Precambrian Shield in Quebec province, the muskeg will be floating and the depth is likely to be 20–30 feet (6.1–9.1 m), and is often more. If, in the photograph or at the actual site, little rock is exposed and vegetation is not interrupted significantly, the slopes will be soil covered. The enclosed muskeg is then shallow, unless localized glacial action induced formation of a kettle hole (Radforth 1962), in which case the peat is deep and floating. In making summaries, it is useful to designate geological conditions with appropriate expressions, for example, kame, esker, drumlin, and moraine (see Glossary). These expressions usually indicate that, where FI occurs, the underlying peat will be deeper and wetter the year round (this is also the case for FI within granitic folds) than for FI bordered by terraces or strands.

In interpretative information, it is useful to indicate clearly whether the muskeg is confined or unconfined. This knowledge helps in deciding how the water resources of the muskeg behave and how they can be controlled. Construction and drainage require an approach and a design that are appropriate to local conditions. FI in confined muskeg is usually associated with either the lagg (an open-water rim at the perimeter of the muskeg) or an open or hidden drainage channel that runs generally in the long axis of the confined area.

Interpretative activity leads directly to recognition of the engineering problem predictable in the light of all applicable facts. Cone penetrometer values and peat density determinations made in accordance with the explanation in the foregoing chapters will reveal strength differences and will assist in arriving at engineering decisions. The mechanical attributes of the peaty material are characteristically different from those for any other cover formula (cf. Table 2.2). Collectively, the values will signify: (1) need for corduroy (or equivalent), (2) design to allow for medium to high primary consolidation with little secondary compression, (3) low elasticity, (4) most rapid rate of deterioration under mechanical stress or drying action, (5) low buoyancy, and (6) very low cohesive strength following remoulding.

It is realized that these interpretative expressions are highly qualitative and – in the range of peat types (Table 2.1) – relative in meaning. Their usefulness lies in the fact that they segregate the mechanical behaviour of this recurring peat type from the behaviour of other common types. The terms are those of first approximation, but they assist the judgment process in determining the design parameters, and augment other advice in the prediction summary. Finally, with the other interpretative features they facilitate identification of specific problems within the major engineering problem.

It is not yet possible to differentiate peat types or categories in quantitative soil mechanics terms of void ratio and permeability. For peat, cohesive strength, elasticity, density, water content, water diffusion potential, and bearing strength may be the significant quantitative expressions for design detail. As yet, these values have to be determined empirically in the field and laboratory once classification is complete. From the total analyses, the primary engineering problem and design parameters are identified.

By the use of prediction summaries, a kind of knowledge that soils engineers will find strange can be co-ordinated. Departure from the procedure used for mineral soils is to be expected; the media are genetically distinct, ordered differently, and behave in contrasting fashion. If these differences are acknowledged, soil mechanics may provide secondary (detailed) information for extended prediction summaries. For example, information of the type shown in Figure 3.16 may be utilized to express qualitative values of peat structure and behaviour.

Prediction for engineering purposes is difficult for the observer using high-altitude airphotography unless there is some indication of what airform pattern means in terms of cover formula. Table 3.2 shows the common cover formula characterizing airform patterns. Classes H, F, and E, in that order, signify increasing intensity

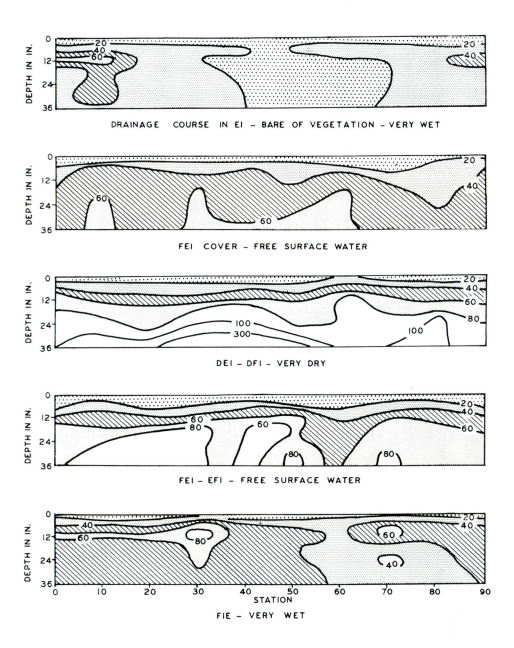

FIGURE 3.16 The association of cone index values in peat beneath related cover formulae showing attempt to express order of homogeneity in peat on the basis of distribution of strength relations (after Radforth and Ashdown 1963)

of grayness in the airphotos. This information, along with any collateral material gleaned from interpretative practice, will help in deciding on the correct formula. The observer should then consult the earlier chapters of this Handbook for peat type distribution and record the results on overlays on the airphotos.

TABLE 3.2
High-altitude airform patterns and
related cover formulae

Marbloid	EH, HE, EI, BEH, BHE
Stipploid	AEH, AEI, AFI, BEI, DFI
Reticuloid	FI, FIE, FEI, BFI
Dermatoid	FI, EI, FEI
Terrazzoid	HE, FI, EH

If the engineering purpose is known, interpretative activity allows the observer to predict basic engineering problems that will occur for given muskeg conditions. Association of proposed types of engineering purpose with muskeg conditions and the problems that arise for the designer are given below. The first of these examples is the summary for the analysis already presented.

Prediction Summary 1

Purpose: Gravel service road – limited light traffic; Precambrian Shield, Northern
   Quebec
Airform pattern: Dermatoid
Cover formula and peat type: FI
Peat category: 2 interrupted by 3 (cf. MacFarlane 1958); 8 and 9 locally present
   immediately to east
Interpretative features:
   (a) Geomorphology of site: confined muskeg; Precambrian trough; shallow
   sand to gravel overburden separates outcrops; floor of deposit uneven with
   mineral deposit (clay locally, silt sand to gravel more widespread, especially
   silt); no subsurface ice after early spring flooding
   (b) Hydrological type: horizontal fen; spring-fed and run-off source; water
   retention not (as yet) due to beaver dam; persistent lagg; seasonal flooding;
   diffuse drainage direction to narrow channel shallow gradient
   (c) Microtopography: featureless (cf. MacFarlane 1958, Table II, pp. 26, 27)
   (d) Macrotopography
Engineering problems and predictions (through interpretation): subsidence with
   unidirectional sliding to west highly possible; no lateral ditching required (or
   desirable); establish off-takes to channel; culvert 6 ft (1.83 m) in diameter at
   south end of traverse or bridge essential; culvert 3 ft (0.91 m) in diameter at
   northwest end of traverse essential; corduroy entire length of traverse (design
   embankment for flotation); do not excavate peat anywhere on traverse; peat
   must not be drained (keep wet to surface)

Prediction Summary 2

Purpose: Proposed traverse by an armoured amphibious, tracked personnel vehicle

Airform pattern: Dermatoid

Cover formula and peat type FI

Peat category: 2 interrupted by 3 (cf. MacFarlane 1958); 8 and 9 locally present immediately to east

Interpretative features:

    (a) Geomorphology of site: as for Summary 1

    (b) Hydrological type

    (c) Microtopography: featureless (cf. MacFarlane 1958, Table II, pp. 26, 27); presence of soft patches 2 ft (0.6 m) in diameter every 20 to 25 ft (6.1 to 7.6 m)

    (d) Macrotopography: as above

Engineering problems and predictions (through interpretation): the vehicle will make a single pass under half load without immobilization in any direction; the soft spots signify limitation for full load pass – otherwise two passes without load, one with load, may be made

Prediction Summary 3

Purpose: Forestry road; heavy transport

Airform pattern: Dermatoid

Cover formula and peat type: EI

Peat category: 13

Interpretative features:

    (a) Geomorphology of site: granitic folds dipping characteristically (cf. Summary 1)

    (b) Hydrological type: fen with water retention (see Summary 1)

    (c) Microtopography: mounds; knolls (ice) spring and autumn; peat depth 16.5 ft (5 m) max. or 11.5 ft (3.5 m) av. overlying sand and some granitic domes

    (d) Macrotopography: as above

Engineering problems and predictions (through interpretation): completely displace peat at laggs (with sand); corduroy not necessary; excavation not necessary; drainage to FI with off-takes; no lateral ditches: will cause subsidence; install berm on loaded side of highway; culverts at laggs (allow for quite temporary thaw flooding); use slash in foundation; do not destroy E cover: end-fill over it

Prediction Summary 4

Purpose: Traverse for same armoured vehicle as used in Summary 2

Airform pattern: Dermatoid

Cover formula and peat type: EI

Peat category: 13

Interpretative features:

(a) Geomorphology of site: granitic folds dipping characteristically (cf. Summary 1)

(b) Hydrological type: fen with water retention (see Summary 1)

(c) Microtopography: mounds; knolls (ice) spring and autumn; peat depth 16.5 ft (5 m) max. or 11.5 ft (3.5 m) av. overlying sand and some granitic domes

(d) Macrotopography: as above

Engineering problems and predictions (through interpretation): 25 passes, full load; 75 passes, no load

Prediction Summary 5

Purpose: Secondary road with asphalt

Airform pattern: Stipploid

Cover formula and peat type: AEI

Peat category: 16

Interpretative features:

(a) Geomorphology of site: continuous muskeg; morainic outwash on old pre-glacial lake-bed; eskers traverse the area at intervals not shorter than 5 miles; peat depth 2 to 5 ft (0.61 to 1.52 m): 5 ft depths rare

(b) Hydrological type: horizontal fen (treed); flooding very rare; not spring fed; retains enough moisture to enable peat to form

(c) Microtopography: mounds (amplitude of height max. 3 ft (0.91 m)) and knolls (ice); few visible deadfalls

(d) Macrotopography

Engineering problems and predictions (through interpretation): excavate peat; establish lateral ditching; off-takes seldom required; interim culverts essential; subsurface silt; clay beneath "dirty" coarse aggregate (shallow – about 1.65 ft (0.5 m))

Prediction Summary 6

Purpose: Traverse for same armoured vehicle used in Summary 2

Airform pattern: Stipploid

Cover formula and peat type: AEI

Peat category: 16

Interpretative features:

(a) Geomorphology of site: as for Summary 5

(b) Hydrological type: as for Summary 5

(c) Microtopography: as for Summary 5

(d) Macrotopography

Engineering problems and predictions (through interpretation): difficult to mount coalescing ice knolls from softer foundation (other cover formula) until mid-summer; track wear severe; smaller trees must be overridden; difficulty reversing; expect down-time; driver fatigue

Prediction Summary 7

Purpose: Construction of forestry service road
Airform pattern: Stipploid
Cover formula and peat type: ADE
Peat category: 15
Interpretative features:

(a) Geomorphology of site: shallow glacial imponding; much local outwash from moraine sources and mixed glacial deposits leaving shallow silts and clays (mostly the former); mixed aggregate beneath.

(b) Hydrological type: horizontal fen muskeg; flood-plain type but in local sites in major area, i.e. multiple flooding for extended period of the spring thaw and run-off; some residual pools in summer and water courses traversing with DFI cover

(c) Microtopography: traps in groups every 100 ft (30.5 m) (approx.); group about 300 ft (91.5 m) wide; single trap shallower than for DFI; peat commonly only 1.65 to 3.3 ft (0.5 to 1 m) deep; many deadfalls

(d) Macrotopography: as above

Engineering problems and predictions (through interpretation): excavate peat; frequent 2 ft (0.61 m) culverts sometimes at regular 200-ft intervals; occasional 5 ft (1.52 m) diameter culverts; ditches deep and close to foot of embankment; construct off-takes from lateral ditches to DFI natural gradients; spring flooding severe; construct gradient accordingly; some berming necessary if fill inadequate

Prediction Summary 8

Purpose: Traverse for same armoured vehicle as used in Summary 2
Airform pattern: Stipploid
Cover formula and peat type: ADE
Peat category: 15
Interpretative features:

(a) Geomorphology of site: as for Summary 7
(b) Hydrological type: as for Summary 7
(c) Microtopography: as for Summary 7
(d) Macrotopography: as for Summary 7

Engineering problems and predictions (through interpretation): track damage; driver fatigue (pitch, roll, and evasion) – deadfalls will be a problem; imponded areas and DFI drainage gradients will present no problem for full load (10 passes) unless track approach is below upper edge of mat in emerging from a drainage course; approach narrow drainage courses on a slight angle if they are no wider than half the width of the vehicle

Prediction Summary 9

Purpose: Construction of service road: petroleum production programme
Airform pattern: Stipploid (stippling uniform – not marginal)
Cover formula and peat type: AFI
Peat category: 10
Interpretative features:
    (a) Geomorphology of site: ancient shallow lake and flood plain with deeper
    50 to 100-ft (15.25 to 30.5 m) wide local depressions; continuous muskeg
    (b) Hydrological type: horizontal fen; combined flood and retention type
    (c) Microtopography: local areas of hummocks (estimated for av. 30 ft
    (9.15 m) across)
    (d) Macrotopography: local imponding (areas wider and deeper by 2 to 3 ft
    (0.61 to 0.91 m) than for ADE condition (cf. Summary 7)) and sluggish
    gradients connecting
Engineering problems and predictions (through interpretation): high maintenance
    costs (re-establishment of grade) – local subsidence; "dishing" at shoulders;
    drain to chain of FI sites; maintain culverts joining FI interruptions of AFI;
    design for interrupted corduroy where A = F in cover formula; no lateral
    ditching, use off-takes at culverts

Prediction Summary 10

Purpose: 10-mile cross-country traverse with fully loaded armoured vehicle,
    same one as used for Summary 2
Airform pattern: Stipploid (stippling uniform – not marginal)
Cover formula and peat type: AFI
Peat category: 10
Interpretative features:
    (a) Geomorphology of site: ancient shallow lake and flood plain with deeper
    50 to 100 ft (15.25 to 30.5 m) wide local depressions; continuous muskeg
    (b) Hydrological type: horizontal fen; combined flood and retention type
    (c) Microtopography: local areas of hummocks (estimated for av. 30 ft
    (9.15 m) across)
    (d) Macrotopography: local imponding (areas wider and deeper by 2 to 3 ft
    (0.61 to 0.91 m) than for ADE conditions (cf. Summary 8)) and sluggish
    gradients connecting
Engineering problems and predictions (through interpretation): local immobiliza-
    tion (where A = F) likely; if route chosen to favour >70 per cent cover for A,
    travel will be optimized (i.e. deviation minimum, subsidence least embarrassing,
    minimum slip, pitch and roll, two passes)

Prediction Summary 11

Purpose: Construction of asphalt road; secondary road for mixed travel – public
    use

Airform pattern: Dermatoid (linear)

Cover formula and peat type: DFI

Peat category: 15, with occasional interruptions of 11, especially to right and left of proposed right-of-way

Interpretative features:

  (a) Geomorphology: glacial outwash configurations; subsequent erosion by paralleling stream drainage; confined muskeg

  (b) Hydrological type: horizontal fen; terraced flooding type

  (c) Microtopography: traps (av. width about 5 ft (1.52 m))

  (d) Macrotopography: lining shallow stream; some open water

Engineering problems and predictions (through interpretation): excavate peat; use bridge – culvert system; install lateral ditching 20 ft (6.1 m) wide and centred 70 ft (21.35 m) from foot of shoulder on downstream side of crossing gradient; displace silt and/or clay under roadbed; high spring flooding and rapid flow around culvert and bridge abutments

Prediction Summary 12

Purpose: Traverse of 50 ft (15.25 m) by fully loaded armoured vehicle, same one as used for Summary 2

Airform pattern: Dermatoid (linear)

Cover formula and peat type: DFI

Peat category: 15, with occasional interruptions of 11, especially to right and left

Interpretative features:

  (a) Geomorphology of site: glacial outwash configurations; subsequent erosion by paralleling stream drainage; confined muskeg

  (b) Hydrological type: horizontal fen; terraced flooding type

  (c) Microtopography: traps (av. width about 5 ft (1.52 m))

  (d) Macrotopography: lining shallow stream; some open water

Engineering problems and predictions (through interpretation): avoid widest traps causing immobilization; cover obstructs vision but do not remove it; force vehicle over traps; 50 passes in one direction if stream bank mounted where D = F locally, otherwise bank cannot be mounted

Prediction Summary 13

Purpose: Secondary road construction (gravel top)

Airform pattern: Stipploid in vicinity of infrequent Marbloid and Reticuloid

Cover formula and peat type: AEH

Peat category: 16

Interpretative features:

  (a) Geomorphology of site: glacial outwash; shallow ridges, domes and fans; continuous muskeg

  (b) Hydrological type: horizontal fen interrupted by aapa fen; sometimes sloping fen; sometimes "bog mire" (blanket and raised types: European systems)

(c) Microtopography: mounds; ice knolls persist to mid-August
(d) Macrotopography
Engineering problems and predictions (through interpretation): site changes character (ice action and thawing); disturbance of properties in foundation; extended season of thaw (subsurface); differences in gravitational water content; construction problems of supply and fill installation; excavation and corduroying not recommended; lateral ditching acceptable; ditches 3 to 5 ft (0.61 to 0.91 m) in width 80 ft (24.4 m) from toe of embankment; shoulder sloughing probable at irregular intervals; differential subsidence and consolidation a function of peat and thaw subsidence sudden; construct in summer

Prediction Summary 14

Purpose: Cross-country traverse of fully loaded armoured vehicle, same one as used for Summary 2
Airform pattern: Stipploid in vicinity of infrequent Marbloid and Reticuloid
Cover formula and peat type: AEH
Peat category: 16
Interpretative features:
    (a) Geomorphology of site: glacial outwash; shallow ridges, domes and fans; continuous muskeg
    (b) Hydrological type: horizontal fen interrupted by aapa fen; sometimes sloping fen; sometimes "bog mire" (blanket and raised types: European systems)
    (c) Microtopography: mounds; ice knolls persist to mid-August
    (d) Macrotopography
Engineering problems and predictions (through interpretation): no immobilization except in late summer; very rough; driver fatigue; route deviation a factor; treed obstacles; deadfalls no problem; immobilization mounting edge of coalescing ice knolls (late summer)

Prediction Summary 15

Purpose: Construction of segment of secondary road (gravel)
Airform pattern: Marbloid 80 per cent; interrupted with Reticuloid, some Polygoid (5000 ft (1525 m) inspection)
Cover formula and peat type: 80 per cent HE; FI depressions; EH edges plateaus
Peat category: 6 interrupted by 1 (where ponding occurs); 3 (low areas FI); 4 (plateau edges)
Interpretative features:
    (a) Geomorphology of site: glacial outwash configurations in permafrost landscape
    (b) Hydrological type: unconfined muskeg; combination horizontal fen (aapa type) and coalescing "bog mire" (raised)
    (c) Microtopography: mounds, ridges

(d) Macrotopography: open and closed ponds; vertical banks; irregular peat plateaus; subterranean drainage; ice is active and on short- to long-term permafrost

Engineering problems and predictions (through interpretation): foundation to be kept frozen all year; differential thawing; do not excavate; search for gravel where H predominates or A interrupts locally; no ditching; off-takes infrequent and short

Prediction Summary 16

Purpose: Traverse with fully loaded armoured vehicle, same one as used for Summary 2

Airform pattern: Marbloid 80 per cent; interrupted with Reticuloid, some Polygoid (5000 ft (1525 m) inspection)

Cover formula and peat type: 80 per cent HE; FI depressions; EH edges plateaus

Peat category: 6 interrupted by 1 (where ponding occurs); 3 (low areas FI); 4 (plateau edges)

Interpretative features:

(a) Geomorphology of site: glacial outwash configurations in permafrost landscape

(b) Hydrological type: unconfined muskeg; combination horizontal fen (aapa type) and coalescing "bog mire" (raised)

(c) Microtopography: mounds, ridges

(d) Macrotopography: open and closed ponds; vertical banks; irregular peat plateaus; subterranean drainage; ice is active and on short- to long-term permafrost

Engineering problems and predictions (through interpretation): slippery slopes (subsurface ice); steep short slopes (3 ft (0.91 m)); shallow narrow clefts cause deviation; serious pitch and roll; bearing strength fails in closed ponds (immobilization); open ponds have steep banks; heavy fuel consumption; driver fatigue; passes: number unpredictable depending on standard of terrain knowledge, otherwise 20 loaded, 50 unloaded

Prediction Summary 17

Purpose: Road construction – secondary road (gravel)

Airform pattern: Marbloid with frequent (40 per cent) interruptions of Dermatoid

Cover formula and peat type: EH

Peat category: 13 interrupted at irregular intervals by 6

Interpretative features:

(a) Geomorphology of site: as for Summaries 15 and 16

(b) Hydrological type: unconfined muskeg; horizontal fen with superimposed "bog mire" (European)

(c) Microtopography: mounds, ice knolls, ridges

(d) Macrotopography: regular peat plateaus; gentle slopes 1 ft in 50 ft (0.3 m in 15.25 m) (frequent)

Engineering problems and predictions (through interpretation): shoulder dishing; irregular consolidation; sliding; avoid parallel ditching; berms locally distributed (i.e. not continuous); difficult to keep frozen; do not disturb cover; no corduroy if berms used; culverts at FI; FEH interruptions; do not cause imponding

Prediction Summary 18

Purpose: Traverse with fully loaded armoured vehicle, same one as used for Summary 2

Airform pattern: Marbloid with frequent (40 per cent) interruptions of Dermatoid

Cover formula and peat type: EH

Peat category: 13 interrupted at irregular intervals by 6

Interpretative features:

(a) Geomorphology of site: as for Summaries 15 and 16

(b) Hydrological type: unconfined muskeg; horizontal fen with superimposed "bog mire" (European)

(c) Microtopography: mounds, ice knolls, ridges

(d) Macrotopography: regular peat plateaus; gentle slopes 1 ft in 50 ft (0.3 m in 15.25 m) (frequent)

Engineering problems and predictions (through interpretation): driver fatigue but not so intensive as for Summary 16; pitch and roll moderate; FI interruptions should be avoided (can be done with little deviation); slippage at near peak of slopes; little downtime

Prediction Summary 19

Purpose: Construction of local road (gravel) near mining community

Airform pattern: Dermatoid (circular to circular-lobed)

Cover formula and peat type: EFI

Peat category: 6

Interpretative features:

(a) Geomorphology of site: drumlin landscape with general morainic features; kettle holes

(b) Hydrological type: horizontal fen spring fed or "bog mire" (flat bog complex: European system); confined muskeg (bog)

(c) Microtopography: mounds and ridges

(d) Macrotopography

Engineering problems and predictions (through interpretation): corduroying necessary (i.e. float road); appropriate construction at lagg (with large diameter culverts); avoid possibility of "folding" or "sinking" along centre line; spring flooding at lagg (edge of EFI) deeper and wider than for DFI conditions (cf.

Summary 11); sinkage will certainly occur without corduroy; avoid parallel ditching; construct off-takes to FI thence to lagg; keep foundation moist; flooding will be mainly at lagg; keep road embankment to 3 ft (0.91 m) of aggregate; settlement is severe but uniform

Prediction Summary 20

Purpose: Traverse using armoured vehicle, same one as used in Summary 2
Airform pattern: Dermatoid (circular to circular-lobed)
Cover formula and peat type: EFI
Peat category: 6
Interpretative features:
(a) Geomorphology of site: drumlin landscape with general morainic features; kettle holes
(b) Hydrological type: horizontal fen spring fed or "bog mire" (flat bog complex: European system); confined muskeg (bog)
(c) Microtopography: mounds and ridges
(d) Macrotopography
Engineering problems and predictions (through interpretation): bellying (immobilization) after two passes with full load; ten passes (no load); failure likely at one track (locally soft, Cover Class I); proper navigation and good operator can avoid sinkage for four passes (full load)

REFERENCES

MACFARLANE, I. C. 1958. Guide to a field description of muskeg (based on the Radforth Classification System). NRC, ACSSM Tech. Memo. 44, rev. ed. 36 pp.

RADFORTH, N. W. 1955. Organic terrain organization from the air (altitudes less than 1000 feet): Handbook No. 1. Defence Res. Board, Dept. Nat. Defence, DR No. 95, Ottawa, 49 pp.

——— 1956a. The application of aerial survey over organic terrain. Proc. Eastern Muskeg Res. Meeting, NRC, ACSSM Tech. Memo. 42, pp. 25–30.

——— 1956b. Muskeg access, with special reference to problems of the petroleum industry. Trans. Can. Int. Min. Metall., Petrol. and Nat. Gas Div., Vol. LIX, pp. 271–277.

——— 1958. Organic terrain organization from the air (altitudes 1000 to 5000 feet): Handbook No. 2. Defence Res. Board, Dept. Nat. Defence, DR No. 124, Ottawa, 23 pp.

——— 1962. Airphoto interpretation of organic terrain for engineering purposes. Trans. Symp. Photo Interpretation, Intern. Archives of Photogrammetry, Vol. XIV, Delft, Holland, pp. 507–513.

RADFORTH, N. W. and K. H. ASHDOWN. 1963. Design in organic terrain and its effect on trafficability. Paper presented at First Regional Meeting, Intern. Soc. for Terrain-Vehicle Systems, Durham, NC.

4 Engineering Characteristics of Peat

Chapter Co-ordinator*
IVAN C. MACFARLANE

4.1 THE STRUCTURAL ASPECT OF PEAT

A soil may be regarded as a system comprised of two or three spatially co-existent phases: a solid phase, a liquid phase, and usually a gas phase. This three-phase concept applies equally well to peat soils and to mineral soils except that for peats the "solid" phase is not always solid but in the microscopic aspect is usually a secondary system of biological entities consisting of cellular structures with contained liquid and/or gas. A recognition of these three phases in peat is basic to an understanding of its engineering characteristics.

In this context, structure refers to the morphology and arrangement of the constituent peat elements, both in the macro- and microscopic aspect. Since peats are comprised of fossilized remains of plant communities, they will contain elements of varying morphology, complexity, and texture. One form of plant remains may be the most predominant visually, another may be secondary in importance, another tertiary, and so on. The structure of peat implies an arrangement of these primary, secondary, tertiary, etc. elements into a certain structural pattern. Figure 4.1 is an example of two such structural patterns. Photomicrograph (a) represents a non-woody fine-fibrous peat and photomicrograph (b) represents an amorphous-granular peat. The microstructure of the two peat types is obviously quite different.

It is this structure of peat, in its various aspects, that affects the retention or expulsion of water in the system, gives it its strength, and ultimately differentiates one peat type from another. This is reflected in Table 2.1 (Chapter 2), which indicates qualitatively the structural spectrum of peat as represented by the Radforth categories 1–17. Category 1 has as its constituents highly disintegrated botanical tissues along with identifiable microfossils; macroscopically its constituents are structurally formless. On the other hand, category 17 is characterized by an open framework of highly preserved fossilized elements, including, with other macro- and micro-constituents, very large woody elements.

*Chapter Consultants: J. I. Adams – Consolidation and Settlement; R. M. Hardy – Shear Strength; K. V. Helenelund – Bearing Capacity; G. P. Williams – Thermal Characteristics.

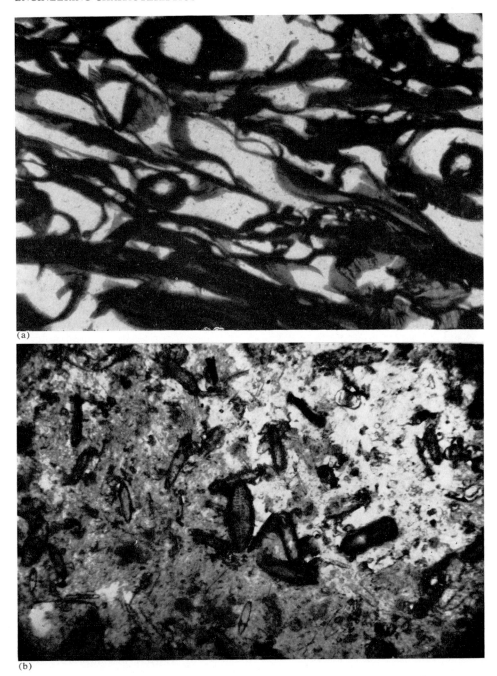

FIGURE 4.1 Photomicrographs of different types of peat structure: (a) a non-woody fine-fibrous peat (magnification 170×), (b) an amorphous-granular peat (magnification 275×)

Virtually no research has been carried out on the correlation of different structural types of peat and their physical, chemical, and strength properties. This is largely due to the fact that no method has yet been devised for expressing peat structure in quantitative terms. MacFarlane (1966) has commenced work on this aspect, supplementing earlier work by Ohira (1962), but much research remains to be done. On a qualitative basis, however, and for the three predominant peat types, relative values of various peat properties are suggested in Table 4.1.

4.2   PHYSICAL AND CHEMICAL CHARACTERISTICS

Certain fundamental physical and chemical properties of peat characterize, to some extent, the quality of that peat relative to engineering purposes. Some of the relevant physical properties are: water (or moisture) content, shrinkage, permeability, void ratio, gas content, unit weight, specific gravity, and per cent ash. Acidity is the chemical property of peat most relevant for engineering purposes, particularly with relation to potential aggressive action on structures.

(1)   Water Content

The amount of water contained in the voids of a peat in its natural state is called the natural water content. Peat has a great capacity for taking up and holding water; this affinity for water is one of the most important characteristics of the material, and in its ability to soak up and hold water, peat acts somewhat like a sponge.

The same drying temperatures as for mineral soils (105–110° C) are normally used in the determination of the water content of peat. There is evidence, however, that such high temperatures may actually char certain types (Goodman and Lee 1962; MacFarlane and Allen 1964; Miyakawa 1959a). Jackson (1958) suggests that the weight loss from drying at temperatures up to 100° C can be attributed to absorbed water, and that the weight loss due to drying at temperatures above 100° C can be attributed to water plus organic matter. Consequently, to reduce the hazard of burning off part of the organic matter, a drying temperature of less than 100° C is recommended, a temperature of 85° C being suggested as appropriate (Goodman and Lee 1962; MacFarlane and Allen 1964), particularly for fibrous peats.

Laboratory determination of the water content of peat is made in much the same way as for mineral soils. To obtain an average value of water content and to avoid local variations, a large representative sample should be used; a suggested minimum size for the sample is one that would contain at least 10 g of dry solids. Drying to constant weight at 85° C will be expedited by the use of a forced-draft drying oven, but even so will require two or three days for a large sample.

Table 4.2 indicates several observed water contents. Except where otherwise noted, however, these values were determined at temperatures of 105–110° C.

TABLE 4.1
Relative values of various peat properties for predominant types

Predominant structural characteristic	Peat property*						
	Water content	Natural permeability	Natural void ratio	Natural unit weight	Shear strength	Tensile strength	Compressibility
Amorphous-granular	3	3	2	1	3	3	2
Fine-fibrous (woody and non-woody)	1	2	1	3	2	2	1
Coarse-fibrous (woody)	2	1	3	2	1	1	3

*On relative scale, 1 is greatest, 3 is least.

TABLE 4.2
Summary of physical properties of peat

Reference	Water content (%)	Permeability (cm/sec)	Void ratio	Unit weight (g/cm³)		Specific gravity	Acidity (pH)	Ash content (%)	Remarks
				Natural	Dry				
Adams (1965)	200–600		3.4			1.62	4.8–6.3	12.2–22.5	Cover: ADI-BEI. Peat Cat. 9–11
	355–425					1.73	6.7	15.9	Cover: F/I. Peat Cat. 2
	330–375					1.65	6.2	12.3	Cover: DFI/B Peat Cat. 6–4
									DFI cover. Peat
Anderson and Haas (1962)	105–470					2.3		47–85	Disturbed ⎫ contaminated by
	100–470							47–95	Undisturbed ⎭ inwash of mineral soil
Anderson and Hemstock (1959)								10–25	@300° C. BEI, FI muskeg
Boelter (1965)		38.1 × 10⁻³							Undecomposed moss peat near surface
		55.8–104 × 10⁻⁵							Woody peats and deep undecomposed moss peat
		0.75 × 10⁻⁵							More dense, decomposed, and herbaceous peats
Brochu and Paré (1964)	300–650		7–11	1.04–1.07	0.147–0.194	1.3–1.75	5–7	15–40	Black fibrous peat (Napierville, Que.)
Casagrande (1966)	250–800	3 × 10⁻⁵	7.1						Near Boston, Mass.
Colley (1950)	485–910		4.6–10.3	0.95	0.20				Fla. peat
Cook (1956)	120–800		2.8–13.1			1.85–2.45		17 (av.)	Vancouver area peat
Farnham (1957)	185–370						6.2	66.6	Amorphous peat ⎫
	300–470						7.5	55.9	Aggregate peat
	1135–2770						3.8	4.5	Moss peat ⎬ Minn.
	530–1260						5.4	15.8	Herbaceous peat
	350–710						7.5	43.9	Aquatic peat
	550–875						6.0	12.5	Woody peat ⎭
Feustel and Byers (1930)	1475–3235					1.39–1.59			Sphagnum peat (Maine)
	1070					1.46			Woody peat (Wash.)
	185–380					1.38–1.55			Heath peat (NC)
	515–1485					1.49–1.61			Sawgrass peat (Fla.)
	350–695					1.11–1.98			Sedimentary peat (Fla.)
Goodman and Lee (1962)	280–320		5.23, 5.48		0.24–0.41	1.51–1.62	4.7–8.6	24.8–59.5	Syracuse, NY ⎫ Water contents
	240–575		4.85, 5.50		0.24–0.26	1.53	6.1–7.1	25–58	East Aurora NY ⎭ obtained at 85° C.

Reference	Water content (%)	Permeability	Void ratio						Remarks
Hanrahan (1952)	340–1465		9–25		0.961–1.01	1.2–1.7			Ireland
Hanrahan (1954)	470–760		Up to 25	0.4 × 10⁻⁴	0.946–1.027	1.1–1.8			Ireland
Hardy and Thomson (1956)				0.064–0.13		1.4			Fibrous peat (Alta.)
Hillis and Brawner (1961)			2–15						Maillardville peat, BC
			5–20						Lulu Island peat, BC
Kapp *et al.* (1966)	Up to 600								Tidal marsh deposits (Newark, NJ)
Lake (1960)	800 av. 1550 av.								Eaglesham, Renfrewshire Dalwellington. Ayrshire
Lake (1961)	400–2000								Concorrat, Dunbartonshire
Lea and Brawner (1963)	500–1500* 2000†	10⁻²–10⁻⁴		0.064	0.881–1.202	1.5–1.6			*Usual range †Occasionally } Vancouver, BC
Lewis (1956)	520 av.				1.046				
Lo and Wilson (1965)	600–700					1.95–2.05	74.5		Amorphous-granular peat (Parry Sound, Ont.)
Lyttle and Driskell (1954)						4.0–5.5			Median range
MacFarlane and Rutka (1962)	105–2780								Variety of peat types in Northern Ont.
Mickleborough (1961)	145–480		3.21–9.86	0.0401–0.344	0.918–1.241	1.6–1.8			Radforth Cat. 9. Deg. of sat. = 80–98 % (Prince Albert Sask.)
Miyakawa (1960)	155–810		6.5–11.4	0.10–0.27	0.95–1.12	1.33–1.91	17–43		Ishikari Peat. Japan. Dried at 60° C
		2–13 × 10⁻³ (horizontal)							Parallel to layer } Kushiro Dist. Japan
		2–7 × 10⁻³ (vertical)							Perpendicular to layer
		5.7–4.86 × 10⁻⁵ (horizontal)							Parallel to layer } Ishikari Dist. Japan
		1.68–7.1 × 10⁻⁵ (vertical)							Perpendicular to layer
Phalen (1961)	205	10⁻⁵–10⁻⁷	4.2	0.114	1.125	1.9			Deg. of sat. = 93.2 %
Radforth (1955)								5–7	Top of peat plateaus } Churchill area
								8–8.5	Bottom of peat plateaus
								8	Flat poorly drained terrain
								4–7	Uneven, better drainage

TABLE 4.2 (*continued*)

Reference	Water content (%)	Permeability (cm/sec)	Void ratio	Unit weight (g/cm³) Natural	Unit weight (g/cm³) Dry	Specific gravity	Acidity (pH)	Ash content (%)	Remarks
Ripley and Leonoff (1961)	100–2100 1000 av.								Radforth Cat. 9, changing to Cat. 4 with depth
Risi et al. (1950–55)							4.9–6.1		Wide variety of Que. peats
Root (1958)	550–650								Fibrous peat (Antioch. Calif.)
Shea (1955)	Usually 500 Often 1000				0.08–0.16		10		Fla. peat
Smith (1950)	150–535								Min. contamination, highly humified
	145–625								Highly humified, no recognizable plant remains
	495–1340								Humified, plant remains still recognizable
	610–1715								Fresh peat
								2.0	Sphagnum moss
								0.6	Cottongrass peat
								14.5	Reed peat
								2.5	Birchwood peat
								10.0	Heath humus
Tessier (1966)	200–890		5–16	0.933–1.028	0.095–0.152	1.9–2.7	6–7	7–13	St. Elie d'Orford, Que.
	300–650		7–11	1.038–1.086	0.148–0.194	1.9–2.68	5–7	15–40	Napierville, Que.
Thompson and Palmer (1951)	240–340		5.1–7.09	0.956–1.134	0.233–0.334	1.795–2.036			Deg. of sat. = 82.5–95.6%
Tveiten (1956)		1–9×10^{-4}							Raw sphagnum peat
		1–5×10^{-6}							Mixed medium peat
		1–3×10^{-6}							Grassy medium peat
		1–2×10^{-6}							Dark grassy peat
Van Mierlo and Den Breeje (1948)		10^{-5}							Holland
Waksman (1942)								2.2–2.5	Moss peat
								6.9	Decomposed heat peat
								22–37.5	Sedge and wood peat
Ward (1948)	800–1000		6–17			1.2–1.5			Wales

The quantity of water held in peats varies considerably, being less for the more decomposed and amorphous types than for the more fibrous types. For pure peats, when saturated, the water content is more than 600 per cent of dry weight, generally varying between 750 and 1500 per cent. A small percentage of inorganic material in the peat will seriously affect the water content value, which will drop sharply as the mineral contamination increases.

(2)   Shrinkage
When peat dries out it shrinks and becomes harder and firmer. Under drought conditions, or because of effective drainage, the surface of a peat bog area may develop shrinkage cracks. In laboratory samples, shrinkage up to 50 per cent of the original volume (Colley 1950) has been observed. When peat is excavated and the material dumped in spoil banks, it can be expected that considerable shrinkage will take place as the peat dries out. Furthermore, when peat is exposed to air, the process of decomposition of the organic matter (which was interrupted by the conditions giving rise to the formation of the deposit) will continue. In due course, the peat will actually disappear under favourable conditions of oxidation and temperature.

Once peat has been dried, it will not again take up as much water as it held prior to drying. Feustel and Byers (1930) observed that, on the average, the amount of water that is taken up again by peat after air drying varies from 55 per cent of that capable of being held in a sphagnum peat to 33 per cent of that capable of being held in a heath peat.

(3)   Permeability
The physical structure and the arrangement of constituent particles in peat greatly affect the size and continuity of the pores and/or capillaries. Such differences result in a wide range of permeabilities in peats. The highly colloidal, amorphous peats tend to discourage permeability by water, whereas the open-meshed fibrous peats are initially quite permeable.

The permeability of peats varies widely, depending upon: (i) the amount of mineral matter present in the peat, (ii) the degree of consolidation, and (iii) the extent of decomposition. The range of values for permeability of a variety of peat types, as observed in the literature, is shown in Table 4.2. In general, the horizontal permeability of peat (particularly of the predominantly fibrous types) is greater than that in the vertical direction (Colley 1950; Miyakawa 1960). Although peat is generally assumed to be quite permeable, the values shown in Table 4.2 indicate that many peat types are, in fact, relatively impermeable. This characteristic has long been recognized in Norway, where peat has been used for the impermeable core of rock dams (Tveiten 1956).

The coefficient of permeability (k) of saturated peat can be measured in the laboratory with the variable head permeameter, in which the flow of water is metered at the inflow end. If h_1 and h_2 are heights above the free water level of water in the inflow tube at times t_1 and t_2 respectively, k is given by the equation

$$k = 2.303 \frac{al}{A(t_2 - t_1)} \log_{10} \frac{h_1}{h_2}, \tag{1}$$

where a and A are the respective cross-sectional areas of the narrow bore tube and the peat sample, and l is the length of the sample. Falling head permeability tests are sometimes carried out in conjunction with consolidation tests on peat.

The permeability of peat does not remain constant under load but decreases markedly for all types of peat. An example of the effect of consolidation on permeability has been given by Hanrahan (1954), who loaded a specimen of partly humified peat with a natural void ratio of 12 and initial permeability of 4×10^{-4} cm per second. After two days under a load of 8 pounds per square inch (0.56 kg per sq cm) the void ratio was reduced to 6.75 and the permeability to 2×10^{-6} cm per second. After seven months under the same load the void ratio was reduced to 4.5 and the permeability to 8×10^{-9} cm per second or 1/50,000 of the initial permeability.

(4)   *Void Ratio*

The void ratio gives an indication of the compressibility of a material; the higher the initial void ratio, the greater the potential compressibility. The natural void ratio of fibrous peats is usually very high, a value as high as 25 having been reported (Hanrahan 1954). The natural void ratio of amorphous peats, on the other hand, is generally quite low, with values as low as 2 having been reported (Hillis and Brawner 1961). The usual range, however, is from 5 to 15 as seen in Table 4.2.

(5)   *Gas Content*

The organic material of which peat is composed and which is submerged beneath the water table is not entirely inert, but undergoes a very slow decomposition, accompanied by the production of marsh gas (methane) with lesser amounts of nitrogen and carbon dioxide. In deposits containing sulphur, hydrogen sulphide is another characteristic product of decomposition. When the water table is lowered, oxidation of the peat occurs, with the subsequent release of carbon dioxide.

The gas content is of considerable theoretical and practical importance. All physical tests are affected by it and, in the field, permeability, rate of consolidation, measurements of pore pressures, etc. are all believed to be substantially affected by the presence of gas (Lea and Brawner 1963).

The effects of gas in peat show up particularly in consolidation results. In a laboratory consolidation test, a large initial compression and an indistinct completion of primary consolidation in time-compression curves reflect the presence of gas (Moran et al. 1958). The effects on the time-compression curves are similar in appearance to secondary compression and some of the large secondary compressions in peats and organic soils may be due to the presence of gas.

The gas (or air) content of peat is difficult to measure and no widely recognized method is yet available. The gas content, when determined, is obtained from data procured during the consolidation test. Moran *et al.* (1958) report a free gas content of peats of 5–10 per cent of the total volume of the soil at atmospheric pressure. Lea and Brawner (1963) similarly report a gas (or air) content of 7–10 per cent for Vancouver, BC peats.

Moran *et al.* (1958) analysed consolidation tests to determine the volumes of free gas typically present in laboratory samples of highly organic soils (not necessarily pure peats). They produced curves of volumes of free gas and of total gas plotted against void ratio for various degrees of saturation. These curves are reproduced in Figure 4.2.

The presence of marsh gas may be a cause for concern when buildings or fills (for parking lots, etc.) are built on a peat bog. The possibility of marsh gas escaping through fissures, holes, or borings and burning upon union with atmospheric oxygen is reported by Moran *et al.* (1958). An examination of the literature and personal inquiries to various agencies involved in construction on peat, however, have failed to reveal a single instance of safety problems arising from the presence of marsh gas. Nevertheless, if in a particular instance the engineer considers it a potential danger, adequate venting to the atmosphere should be provided, possibly by using a coarse granular fill.

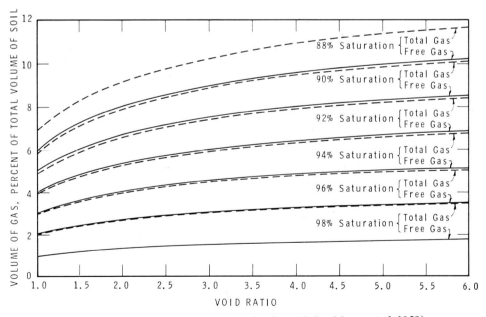

FIGURE 4.2   Volume of gas in a laboratory sample of peat (after Moran *et al.* 1958)

(6)   *Unit Weight*

The natural unit weight depends upon the water and organic content of peat; for a saturated pure peat this weight is usually about the same as that of water. Higher unit weights are usually associated with substantial inorganic content. Natural unit weights have been observed to range from 25.0 pounds per cubic foot (0.4 g per cu cm) for a moss peat to 75.0 pounds per cubic foot (1.2 g per cu cm) for an amorphous peat. Dry unit weights range from 5 pounds per cubic foot (0.08 g per cu cm) to 20 pounds per cubic foot (0.32 g per cu cm), the latter value representing considerable mineral soil contamination. A summary of observed values is given in Table 4.2.

The natural unit weight can be determined by the liquid displacement method, similar to that for inorganic soils. Alternatively, a chunk sample or a sample obtained in a large diameter, thin-walled sampler can be carefully measured and weighed and the unit weight determined in this manner.

The dry unit weight (γ_d) may be determined from the formula

$$\gamma_d = \frac{G}{1 + e} \, \gamma_w, \tag{2}$$

where G is the specific gravity of soil solids, e is the void ratio, and γ_w is the unit weight of water.

(7)   *Specific Gravity*

The specific gravity of the solids in peat is greater than 1. In the literature (see Table 4.2) the range of specific gravities is from 1.1 to 2.5, with the average being about 1.5 or 1.6. Specific gravity values greater than 2.0 indicate a peat with a considerable degree of mineral soil contamination.

Accurate determination of the specific gravity is somewhat difficult owing to the presence of entrapped air or gases. This can be overcome by pulverizing the oven-dried peat with a mortar and pestle. The specific gravity can then be measured in the usual way by the liquid displacement method. Water can be used, but more accurate results are achieved by using kerosene with a known specific gravity.

A satisfactory experimental procedure for determining the specific gravity of peat solids is outlined by Akroyd (1957, pp. 46–54). Essentially it involves placing the pulverized peat sample in a flask or density bottle, covering it with de-aired, filtered kerosene, and applying a high vacuum until air bubbles cease to be emitted from the sample. The container is then filled with kerosene and permitted to reach a constant temperature. The specific gravity may be calculated from the equation

$$G = \frac{\text{weight of dry soil}}{\text{weight of kerosene displaced}} \times \text{specific gravity of kerosene.} \tag{3}$$

An approximate and more rapid method of determining the specific gravity of peat has been suggested by Cook (1956) and Lea and Brawner (1963). They used the concept of ash content, assuming that the ash is composed of clay minerals with a specific gravity of 2.7 and that the organic material (liquid and wood) burnt off during the firing has a specific gravity of 1.5. The average specific gravity of the peat solids is then calculated from the equation

$$G = (1 - A_c)\ 1.5 + 2.7\ A_c,\tag{4}$$

where A_c is the ash content.

Doyle (1963) has estimated, however, that these assumptions can result in an error of 18 per cent in an extreme case.

(8) Acidity

Usually peat has an acidic reaction caused by the presence of carbon dioxide and humic acid arising from its decay. Some of the simpler organic acids (such as acetic acids) may also be produced in small quantities. Peaty waters, which are practically free from salts, generally show pH values of 4–7 (Lea 1956), although values as low as 2.0 and as high as 8.0 have been reported. Table 4.2 shows observed values for several peats. The acidity of peat waters fluctuates with the seasons and weather conditions and is usually highest after a heavy rain following a warm dry period. This characteristic is important because peats and peaty waters can be potentially corrosive to concrete and steel structures (see Section 7.7). Soil and water pH can be simply measured in the field with special litmus paper available commercially for this purpose. Alternatively, various small portable battery-operated pH meters are commercially available for field measurements, as well as much more sophisticated and accurate models for laboratory operation.

(9) Ash Content, Organic Content

The organic material of peat is generally combustible carbonaceous matter, whereas the mineral constituent – whether part of the plant growth or extraneous matter – is incombustible and ash forming. Peat that is mainly free of extraneous mineral matter may have an ash content as low as 2 per cent (see Table 4.2) based on oven-dry weight. From this minimum, every gradation is found up to the point at which the soil is no longer considered to be predominantly peat, but can be classed as an organic clay or silt.

The ash content of peat is determined by firing an oven-dried peat sample in a muffle furnace at a temperature of 800–900° C for three hours or until the peat has obviously been reduced to an ash. Alternatively, the peat sample can be fired in a crucible over a bunsen burner, in which case extreme care must be taken that the ash is not carried away by the hot air currents. With either method, the ash content is determined from the equation

$$A_c(\%) = \frac{\text{weight of ash or residue}}{\text{dry weight of sample}} \times 100.\tag{5}$$

The organic content has a considerable effect on the physical and mechanical properties of peat. In general, the greater the organic content the greater the water content, void ratio, and compressibility of the peat.

Accurate determination of the organic content of peat is a rather long and laborious procedure which involves treating the peat sample with chromic acid in hot sulphuric acid, then quantitatively determining, by titrating against a standard ferrous ammonium sulphate solution, the excess chromic acid that remains after oxidation of the carbon (MacFarlane and Allen 1964). Consequently, in standard engineering practice the ash content is determined as already described and the per cent organic content is considered to be $100 - A_c$. In the firing process, however, more than organic carbon is burned off. This can, therefore, be considered only an approximate method of determining the organic content, and can err from 5 to 15 per cent.

(10)   *Consistency Limits*

The consistency, or Atterberg, limits indicate the range of water content in which a soil may be considered as a fluid, plastic, or solid.

Peats generally do not lend themselves to the carrying out of standard consistency tests. The presence of fibres in most peats makes both the liquid and plastic limit tests difficult, if not impossible. For amorphous-granular peats, however, such tests may be practicable, although reports in the literature are extremely rare. Lewis (1956) carried out Atterberg limit tests on an English peat with an average water content of 520 per cent. He found the liquid limit to be 670 per cent and the plasticity index to be 230 per cent.

Casagrande (1966) suggests that peat ranges from low plasticity for thoroughly weathered deposits to non-plasticity for highly fibrous deposits. He showed a plasticity chart for peats (reproduced in Fig. 4.3) in which all points are located below the A-line (see Glossary). The two heavy lines and the encircled area represent an average of numerous test results for a thoroughly weathered peat in northern Germany and slightly fibrous peat deposits in the United States, with liquid limits ranging between approximately 300 and 1000 per cent. Casagrande suggests that, with increase in fibrosity, the points on the chart move down and to the left, as indicated by the two dashed lines.

4.3   THERMAL CHARACTERISTICS

Peats have a low coefficient of heat conduction, the specific value of which depends upon the porosity and water content of the peat. In any condition, peat acts as a good insulator (Brown 1963). For example, the organic mantle on the ground surface forms a natural insulator for permafrost from the thawing effects of the summer. This concept has long been recognized qualitatively and construction techniques in permafrost areas involve disturbing the moss and peat as little as possible (see Section 7.6). Even early attempts at a quantitative appraisal of the insulation quality of peat recognized the abnormally high water-retention capacity of peat and its varying thermal conductivity in the thawed and frozen states (Muller 1945).

This insulating value of peat has been recognized for many years in Norway, where it has been used beneath the bearing layer in railway embankments to protect against frost heaving (Skaven-Haug 1959). In the Soviet Union, also, peat is used in the production of insulating board for the housing industry (New Scientist 1964).

To calculate the rate of freezing and rate of thawing under changing water content, information is needed on the thermal properties of peat soils, and on how they change with changing water content (Williams 1966). The thermal properties that determine heat transfer in a soil and, hence, depth of freezing or thawing are: (1) volumetric specific heat (or heat capacity) (C_v), (2) latent heat (L), (3) thermal conductivity (K), (4) thermal diffusivity (a), which is defined by the ratio K/C_v, and (5) conductive capacity, which is defined by the product $C_v \sqrt{a}$.

(1) Volumetric Specific Heat
The specific heat per unit volume of soil can be found by adding the heat capacities of volume fractions of solid material, water (or ice), and air. If X_s, X_w, and X_a

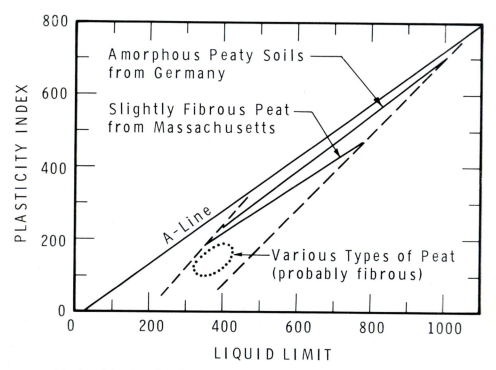

FIGURE 4.3   Plasticity chart for peat (after Casagrande 1966)

denote the volume fractions of solid material, water, and air respectively, then

$$C_v = X_sC_s + X_wC_w + X_aC_a, \tag{6}$$

where C_s, C_w, and C_a are the specific heat of the different soil constituents. The third term can be neglected (C_a is small), so that

$$C_v = X_sC_s + X_wC_w. \tag{7}$$

Assuming a value of C_s for peat soil as 0.6 calories per cu cm per °C and a value of C_w as 1.0 calories per cu cm per °C, the heat capacity of the soil is $C_v = 0.6\,X_s + 1.0\,X_w$ calories per cu cm per °C. Table 4.3 gives calculated values of C_v for different values of X_s and X_w.

TABLE 4.3
Calculated values of volumetric
specific heat for unfrozen peat

	X_s		
X_w	0.10	0.20	0.30
0.5	0.56	0.62	0.68
0.6	0.66	0.72	0.78
0.7	0.76	0.82	0.88
0.8	0.86	0.92	0.98
0.9	0.96	1.02	1.08

TABLE 4.4
Calculated values of volumetric
specific heat for frozen peat

	X_s		
X_{ice}	0.10	0.20	0.30
0.5	0.29	0.34	0.39
0.6	0.33	0.38	0.44
0.7	0.37	0.42	0.48
0.8	0.42	0.46	0.52
0.9	0.46	0.51	0.56

For frozen organic soil, $C_{ice} = 0.45$ calories per cu cm per °C and C_v frozen $= 0.6\,X_s + 0.45\,X_{ice}$. Table 4.4 gives calculated values of C_v frozen (calories per cu cm per °C).

(2)   *Volumetric Latent Heat of Fusion*
The volumetric latent heat of fusion, L, is easily calculated if X_w is known. For example, if $X_w = 0.5$, then L equals 0.5(79.7) or 39.8 calories per cu cm and at $X_w = 0.9$, L equals 72 calories per cu cm.
A column of soil, 45 cm deep with a cross-sectional area of 1 sq cm would have

a total L of 1790 calories at $X_w = 0.5$, and 3240 calories at $X_w = 0.9$. The volumetric specific heat for the same volume would be only about 25 calories (at $X_w = 0.5$ and $X_s = 0.1$). Since the volumetric specific heat is much less, it is often neglected in depth-of-freezing and rate-of-thaw formulae (Kersten 1959).

(3) Thermal Conductivity

De Vries (1963) has developed an equation for calculating the thermal conductivity of soil, including peat, if X_a, X_w, and X_s are known or can be estimated. He assumes that for peat the soil particles are long cylinders of circular cross-section, that the air-filled pores are also long cylinders, and that, for values of $X_a > 0.5$, water is a continuous medium. De Vries' procedure, based on these assumptions, can now be used to calculate values of K (millicalories per cm per sec per °C) for unfrozen and frozen peat soil (Tables 4.5, 4.6).

TABLE 4.5
Calculated values of thermal conductivity for unfrozen peat

	X_s		
X_w	0.10	0.20	0.30
0.5	0.66	0.72	0.79
0.6	0.79	0.86	0.93
0.7	0.94	1.01	1.09
0.8	1.08	1.17	1.25
0.9	1.25	1.35	1.45

TABLE 4.6
Calculated values of thermal conductivity for frozen peat

	X_s		
X_{ice}	0.10	0.20	0.30
0.5	2.2	2.3	2.4
0.6	2.7	2.8	2.9
0.7	3.3	3.4	3.5
0.8	3.9	4.1	4.3
0.9	4.6	4.8	5.0

If the volumetric water content is doubled, the thermal conductivity is more than doubled. If the percentage of solid material (X_s) is doubled or tripled, there is only a slight increase in the thermal conductivity. The thermal conductivity of the frozen soil is about four times that of unfrozen soil.

Thermal conductivity values are useful not only in heat exchange considerations but also in estimating the insulating value of peat used in construction to retard frost action, as an insulating board, etc.

(4) *Thermal Diffusivity*

The calculated values of thermal conductivity and specific heat are used to calculate values of the thermal diffusivity (sq cm per second), for a range of X_w and X_s. The thermal diffusivity of frozen and unfrozen peat soils is compared in Table 4.7.

To slide rule accuracy, there is no change in a for unfrozen soil, as X_s changes from 0.1 to 0.3, and only a small change in a as X_w changes from 0.5 to 0.9. For frozen soil there is a slight decrease in a, as X_s increases, and a small increase as X_w increases. The thermal diffusivity of the frozen soil is 8–10 times that of the unfrozen soil.

TABLE 4.7
Calculated values of thermal
diffusivity for unfrozen and
frozen peat

	X_s		
X_w	0.10	0.20	0.30
0.5	0.0011	0.0011	0.0011
0.9	0.0013	0.0013	0.0013
	X_s		
X_{ice}	0.10	0.20	0.30
0.5	0.008	0.007	0.006
0.9	0.010	0.009	0.009

TABLE 4.8
Calculated values of conductive
capacity for unfrozen and
frozen peat

	X_s		
X_w	0.10	0.20	0.30
0.5	0.019	0.021	0.023
0.9	0.034	0.036	0.039
	X_s		
X_{ice}	0.10	0.20	0.30
0.5	0.025	0.028	0.031
0.9	0.046	0.049	0.053

(5) *Conductive Capacity*

The calculated values of C_v and K are also used to calculate values of the conductive capacity $C_v \sqrt{a}$ (calories per sq cm per °C per $\sqrt{\text{second}}$). The conductive capacities of frozen and unfrozen peat soils are compared in Table 4.8.

The conductive capacity of the frozen soil is only about 30 per cent higher than that of the unfrozen soil. This indicates that the rate of heat transfer into frozen soil will be only about 30 per cent greater than the rate of transfer in unfrozen soil under similar surface temperature conditions. In Table 4.9 these results are compared with values reported in the literature (Priestley 1959).

TABLE 4.9
Conductive capacity for various materials

Material	$C_v \sqrt{a}$
New snow	0.002
Old snow	0.012
Organic soil	0.04
Ice	0.05
Peat soil	
Unfrozen	0.019–0.039
Frozen	0.046–0.053

The value of $C_v \sqrt{a}$ for frozen peat soil of high water content approaches that of ice. The value for unfrozen peat soil is an order of magnitude larger than that of new snow, and double that of old snow.

4.4 STRESS-DEFORMATION CHARACTERISTICS

(1) Shear Strength

The great majority of practical engineering problems involving peat, if they are to be rationally analysed, require a knowledge of its shear strength characteristics. Problems of trafficability are primarily concerned with the surface mat of live material, but those involving the stability of road embankments and dykes are concerned with the strength throughout the whole depth of the peat deposit, and its variation under loading. This section deals primarily with the shear strength properties of the peat below the surface mat of living vegetation.

The "spongy" nature of peat is one of its obvious characteristics. Technically this means that large deformations occur as the peat develops its inherent resistance to applied loads. In fact, the deformations are extraordinarily large as compared to almost all inorganic soils with which engineering practice is usually concerned.

A second property of peat that is unusual, in comparison with the inorganic soils that are normally encountered in engineering practice, is the range of natural water content over which peat exists, as low as 50 per cent but as high as 2000 per cent, based on the dry weight of its solid phase. This means that its natural water content may be from 10 to 100 times greater than the natural water content normally found in inorganic soils.

In addition, peat usually exhibits a high degree of anisotropy with respect to permeability, which is much higher in the horizontal direction in the deposit than in the vertical direction. Moreover, there are reasons to doubt that the solid phase in peat can indeed be considered as inert and incompressible, for practical purposes, as is assumed for the inorganic soils.

For these reasons early research concerning the engineering characteristics of peats was directed to the problem of whether they could be assessed in terms of the current concepts of soil mechanics which were developed primarily for inorganic soils. Since about 1952 extensive data have been accumulated from both laboratory and field tests which show that the normal principles of soil mechanics are applicable, for practical purposes, to peat soils, but some anomalies do exist.

Qualitatively, the shear strength of peat varies inversely with its water content and directly with its ash content and degree of deformation in compression (Wyld 1956). Generally, there is an increase in strength with depth, but the available field data are not completely consistent in confirming this trend. Higher quality peat, as evidenced by heavier tree growth, appears to show a substantial increase in strength with depth, but poorer quality peat may show little, if any, increase (Anderson and Hemstock 1959; Lea 1962). All published test results show that peat exhibits a relatively high degree of sensitivity, as measured by the ratio of maximum shear strength developed to that finally remaining after continued straining. Sensitivity values ranging from 1.5 to as high as 10 have been reported (MacFarlane 1962). All published test results also indicate that peat shows an extraordinarily high increase in strength as it consolidates.

For purposes of engineering design, numerical values must be assigned to the shear strength of peat. In the first instance, the *in situ* strength must be estimated. A large body of strength data have been published since 1954, based on *in situ* strength measurements using shear vanes and on a variety of types of laboratory tests, with the data covering all the more common types of peat (Assoc. Comm. on Soil and Snow Mech. 1954–1962; Hanrahan 1954). *In situ* shear strengths have been reported in the range from 30 pounds per square foot (146.5 kg per sq m) to 2000 pounds per square foot (9765 kg per sq m). At peat depths of less than 5 or 6 feet (1.52 or 1.83 m), however, the *in situ* shear strength will generally be within the range of 100 to 400 pounds per square foot (488 to 1953 kg per sq m). Refinements to these figures for a particular project require *in situ* shear vane tests on the actual site. For a fuller discussion of the shear vane test and for additional test results, the reader is referred to Section 5.2(1) and to Table 5.4.

In the great majority of practical situations, however, values are also required for parameters that will reflect the change in shear strength with variations in stress environment and loading history for the peat. An integral part of such considerations is the compressibility characteristics of peat. Two separate approaches have been used. The earlier efforts were directed to the establishment of relations between shear strength and water content, with these in turn being related to compressibility characteristics as indicated by compression index values (Hanrahan 1954; Lea and Brawner 1959). The preferable approach, and the one that predominates in more recent publications, is that of assessing the shear strength in terms of an angle of internal friction. This brings the problem into the realm of the conventional approach used in soil mechanics for inorganic soils.

Conventional practice in soil mechanics uses the Coulomb–Hvorslev equations, which give the shear strength in terms of a cohesion component and a frictional component expressed in terms of the angle of internal friction for the soil and the normal stress acting on the soil element. The classical Coulomb equation states

$$\tau_f = c_u + \sigma \tan \phi_u, \tag{8}$$

where τ_f is the shear strength, c_u is the apparent cohesion, σ is the normal stress on the soil element, and ϕ_u is the apparent angle of internal friction. Equation (8) is known as the shear strength relationship in terms of total stress. The difficulty with equation (8) is that both c_u and ϕ_u are dependent on the test conditions, and therefore are not unique soil parameters. The Coulomb–Hvorslev relation modifies equation (8) to correct for these deficiencies, and states that

$$\tau_f = c' + \sigma' \tan \phi', \tag{9}$$

where c', σ', and ϕ' are the effective cohesion, the effective normal stress, and the effective angle of internal friction, respectively. Values for c' and ϕ' can usually be evaluated by tests as unique parameters for the particular soil.

Early tests for the shear strength of peat suggested that its strength was largely cohesive (Hanrahan 1954), but more recent research has shown conclusively that it is essentially a frictional material and that it behaves closely in accordance with the principles of effective stress (Adams 1965). Thus, for practical purposes in problems involving peat, equations (8) and (9) can be simplified to

$$\tau_f = \sigma \tan \phi_u \tag{10}$$

and

$$\tau_f = \sigma' \tan \phi'. \tag{11}$$

Equations (10) and (11) are both useful in solving practical problems with peat. Where the increase in strength with load is measured by *in situ* tests, the results will evaluate ϕ_u in equation (10). Laboratory tests can be used to evaluate either ϕ_u in equation (10) or ϕ' in equation (11), depending on the test procedures.

There is now an extensive body of test data indicating that the values of both ϕ_u and ϕ' are exceptionally high as compared to those for most inorganic soils. Values for ϕ' as high as 50° have been measured (Adams 1961), whereas values of 25 to 30° have been deduced for ϕ_u (Lea and Brawner 1959). For many practical problems, the shear strength of the peat is not the governing factor in the design, which is more likely to be governed by the deformation characteristics of the peat. In such cases, values for ϕ' or ϕ_u can usually be estimated with sufficient accuracy, taking into account that a relatively high factor of safety should be used. For cases where higher precision is desirable, either *in situ* or laboratory tests are necessary.

Stability analyses involving peat soils are usually made by the conventional methods of comparing computed stresses on potential failure arcs, or alternatively on the base of potential sliding blocks, with the estimated strength of the peat. The major difference from the conventional procedure for inorganic soils is in the assumption that the total shear stress must be resisted by the shear resistance on the portion of the failure surface passing through the peat. The reason for this is that it must be anticipated that rupture will occur in inorganic soils before these soils can deform to the extent necessary for the peat to develop its ultimate strength.

Recent work (Adams 1961), however, has established that the K_0 value for peat is exceptionally low and that it decreases with increasing effective consolidating pressure. For normal inorganic soils, K_0 usually has values ranging from 0.4 to 0.8. Adams (1961) reports test results in which K_0 had an initial value under low load of the order of 0.5, but decreased with increasing consolidating pressure and deformation to a value of the order of 0.175. This finding has important implications in stability analyses involving bearing capacity and the development of shearing resistance on failure surfaces. It suggests that these latter methods may give results that are unsafe, and that stability analyses should be extended to include an analysis of the shear stresses induced in the peat, assuming K_0 values approaching zero.

This finding of the unusually low K_0 values for peat appears to explain a number of failures that have occurred in embankments on muskeg despite their being subjected to engineering design and control of construction.

(2) *Tensile Strength of Peat*

The reliability of vane tests for determining the shear strength of fibrous peats is questioned by Helenelund (1967), even though the technique is widely used by various investigators, as reported elsewhere in this Handbook (Section 5.2(1)). In view of the complex structure of fibrous peat, in any study of its strength characteristics there is a valid argument for investigating its strength and deformability in tension as well as in compression and shear. Tension tests are commonly used, of course, in the evaluation of the strength properties of many engineering materials, including inorganic soils (Haefeli 1944).

Helenelund (1967) developed a method of measuring the tensile strength of fibrous peat, utilizing a lightweight aluminum tension box. This box is divided into two parts, one part fixed and the other part resting on ball bearings and moving in a horizontal direction (Fig. 4.4). The large peat sample (12 by 8 by 8 inches or 30.5 by 20.3 by 20.3 cm) is fixed to the tension box through four nail plates, two at the top and two at the bottom of the sample. As the half boxes are pulled apart, the soil specimen is stressed in tension and ultimately a vertical failure plane develops in the middle of the specimen (Fig. 4.5).

Although this method of determining the tensile strength of fibrous peat is still in the experimental stage, the results are quite significant (Helenelund 1967). The horizontal tensile strength of a sphagnum peat varied between 74.2 pounds per square foot (362 kg per sq m) and 233 pounds per square foot (1138 kg per sq m), the mean value for several tests carried out under field conditions being 144 pounds per square foot (703 kg per sq m). For controlled condition laboratory

FIGURE 4.4   Tension box apparatus

FIGURE 4.5   Failure plane in peat sample

tests, the corresponding limits were 82.5 pounds per square foot (402 kg per sq m) and 206 pounds per square foot (1005 kg per sq m), with the mean tensile strength being 136 pounds per square foot (664 kg per sq m), very close to that for the field conditions. A few tests were performed on a *Carex* peat, for which the horizontal tensile strength ranged between 124 and 206 pounds per square foot (605 to 1005 kg per sq m).

Simple beam tests have also been done by using the upper layer of fibrous peat. It has been possible to carry out tests on beams of sphagnum peat of a length/height ratio (l/h) up to 3–4, but the distance between the supports of a centric loaded, simply supported beam should generally be less than 2.5–3.0 h to avoid failure due to the weight of the beam itself. Helenelund (1967) utilized beams 40 by 15 by 15 cm (15.75 by 5.9 by 5.9 inches) spanning 30 cm (11.8 inches) (Fig. 4.6). Assuming that peat behaves elastically, its mean flexural strength was calculated to be about 2.2 times the mean tensile strength measured in the tension box apparatus. Assuming that the peat behaves plastically, the flexural strength is smaller, its mean value being about 1.5 times the mean tensile strength.

The mean shear strength of the sphagnum peat, determined with a four-bladed vane with a diameter of 10 cm (4 inches) and a height/diameter ratio of 2, was 288 pounds per square foot (1.4 ton per sq m) or twice the mean tensile strength. The undrained shear strength is, therefore, of the same order as the flexural strength calculated on the basis of elasticity.

(3)   *Strain Characteristics*

When peat is stressed – whether in tension, compression, or in shear – very large deformations result, compared to those for most inorganic soils. Very little fundamental research has been carried out on the stress–strain relationship in various peat types. Schroeder and Wilson (1962) studied an amorphous-granular peat and concluded that it was a pseudo-plastic material. Tressider (1958) stated that, when loads on peat are small or rapidly applied, both elastic and viscous properties are readily noticeable, elastic effects being particularly associated with the more fibrous and less humified varieties of peat.

Maximum shear strength in peat is developed only after an extremely high degree of deformation (Hardy and Thomson 1956; Tressider 1958). Von Moos (1961) reports that, in some shear vane tests, the torque did not reach its maximal value until an angle of rotation of 180–270° was achieved. In tests with various types of shear vanes in fibrous peats, Helenelund (1967) did not find such extremes but reports that for a two-bladed vane the torque reached its maximal value at an angle of rotation of 80–100°, at 40–75° for a four-bladed vane, and at 30–60° for a six-bladed vane.

Probably the case of peat loaded in compression provides the most important application of the large strains. Hanrahan (1964) indicates that some elastic settlement takes place immediately upon application of the load. No change of water content is involved and, consequently, the strength of the peat remains

FIGURE 4.6   Beam tests of peat

unaltered. If appropriate values of the elastic constants, Young's Modulus (E) and Poisson's Ratio (v), are available for the peat, the amount of elastic settlement of a strip load of width B and unit load p may be calculated from the formula of Steinbrenner (1934):

$$S_e = p(B/E)[(1 - v^2)F_1 + (1 - v - 2v^2)F_2], \qquad (12)$$

where S_e is the elastic settlement and F_1 and F_2 are factors available from graphs developed by Steinbrenner.

The evaluation of reliable elastic constants is by no means straightforward, however, and Hanrahan (1964) proposes the following simple empirical formula (based on experimental data obtained with an Irish peat) for estimating this type of settlement:

$$S_e = 0.24 \, B \, p/\tau_f, \qquad\qquad (13)$$

where S_e is the elastic settlement (inches), B the width of loaded area (feet), p the unit load on loaded area (pounds per square inch), τ_f the vane shear strength (pounds per square inch).

The formula is based on the assumption that Young's Modulus is proportional to the vane shear strength. Also, the formula properly applies only when the depth of the peat layer is six times the width of the foundation, which was the ratio obtaining for the original laboratory measurements. For other ratios, a correction should be made in accordance with the standard methods of elastic theory.

Another mode of deformation under a load is settlement accompanied by a gradual lateral expulsion of the soil from beneath the footing. This lateral creep results essentially from the compressible soil being subjected to appreciable shear stresses. It may give rise to upheaval and ultimately to partial or complete rotational failure. Despite its importance, this type of settlement has been given remarkably little attention in the literature. Hanrahan (1964) draws attention to the problem and suggests that lateral creep is closely akin to secondary consolidation (or compression). He relates the magnitude of the creep settlement to the vane shear strength of the peat. Just how much of the creep settlement is due to triaxial shear stresses and how much of it is actually due to "consolidation creep" (Burmister 1959) is rather difficult to evaluate. Jumikis (1962, p. 467) suggests a formula for determining (approximately) the magnitude of this settlement for various shapes of foundation:

$$\Delta H = c_4 \, p \, (P/A), \qquad\qquad (14)$$

where ΔH is the settlement, p is the contact pressure, and P is the perimeter of the loaded area A. The formula contains a constant c_4 (a coefficient containing the effect of the internal friction of the soil, and different for different soils), however, the value of which is very difficult to ascertain. Jumikis advises that it be evaluated by tests on the particular soil in question. It is apparent, therefore, that there is need for some research on this aspect of the settlement problem.

In a practical example of water storage tanks on a preloaded marshy area containing peaty organic silt, Kapp et al. (1966) report that the total settlement was 2.2 feet (0.67 m) greater than that initially calculated. The initial investigation, prior to the installation of the tanks, considered the settlement resulting only from consolidation and did not take into account settlement due to triaxial stresses. The major difference between theoretical and observed settlement is attributed to strains due to triaxial stresses. In their analysis of the situation Kapp *et al.* made an attempt to estimate shear settlements by methods similar to those described by Skempton and Bjerrum (1957) as well as Lambe (1964). Triaxial tests showed that the *A* value close to failure ranged from 0.75 to 1.17; the field value was observed to be very close to 1.0.

In analysing the strains due to the shear stresses, and considering the increase in strength due to the initial lift (a blanket load), Kapp *et al.* found that the middle 9 feet (2.75 m) of the peaty organic silt layer under the centre of the tank appeared to be stressed beyond its ultimate strength. To estimate the stresses, they used the stress distribution based on the Boussinesq theory (Jurgenson 1934; Gray 1936). Since shear failure did not occur, they conjectured that either an increase in lateral stresses must have occurred when the soil was strained beyond its ultimate strength, or the soil was stronger than the results of triaxial tests seemed to demonstrate. Another obvious possibility was that the stresses were not best described by this theory. Considering only strains due to stresses below the ultimate strength of the soil, Kapp *et al.* accounted for all but 0.28 foot (8.5 cm) of the observed settlement. It was estimated that an additional average strain of 3.1 per cent for the middle 9 feet (2.75 m) of the layer would be required to account for the total observed settlement.

On the basis of this analysis, supported by piezometer and settlement plate readings from under the tanks, Kapp *et al.* conclude that "the observed settlement is attributed to a combination of rebound recompression, remaining primary (not eliminated by the preload), secondary compression, and creep due to triaxial stresses. During the early stages the post-construction settlement is considered to be primarily due to recompression, and perhaps also due to some remaining primary, and creep due to triaxial stresses. The long-term settlements should be primarily due to secondary compression."

Burmister (1959) performed creep tests on peats and organic silts. His tests demonstrated that the undrained shear strength of these soils is strain-rate dependent, the slower the rate of strain the lower the shear strength. The shear strength of peats and organic silts remained relatively constant when tested at a strain rate of 0.00001 of the normal strain rate of 0.55 per cent per minute. At this very low strain rate, the lowest shear strength obtained was 66 per cent of the value obtained when performing the tests at the normal strain rate for the organic silts. It was somewhat higher, however, for the peats.

(4) Bearing Capacity

Bearing capacity considerations are important both in the construction of embankments on the muskeg surface and in vehicle mobility. The former aspect is discussed to some extent in Section 4(1) of this Chapter and in Chapter 6, the latter is discussed in Chapter 8. In the present section the bearing capacity of the surface layer (undecomposed or slightly decomposed peat) is considered, together with the estimation of bearing capacity based on experience.

A   Bearing Capacity of the Surface Layer

The bearing capacity of peat is generally estimated by using methods of calculation similar to those applied to mineral soils – that is, the soil is assumed to fail in shear. In one of the earliest investigations on the strength and bearing capacity of peat, Smith (1950) made the assumption that the ultimate bearing capacity (p_u) is twice the critical load (p_c), giving a maximum shear stress (τ_{max}) equal to the shear strength (τ_f) of the peat. He used the theory of elasticity, which for a strip load on the soil surface gives

$$\tau_{max} = p/\pi, \tag{15}$$

where p is the uniformly distributed surface load. This gives an ultimate bearing capacity of

$$p_u = 2p_c = 2\pi\,\tau_f, \tag{16}$$

compared with

$$p_u = (\pi + 2)\,\tau_f \tag{17}$$

according to Prandtl's solution based on the theory of plasticity.

Because of the high compressibility of peat, a load applied to the muskeg surface causes considerable initial settlement, the magnitude of which is dependent upon the thickness of the peat deposit and the width of the loaded area. In many cases the settlement is so great that the muskeg can be considered to have failed in compression. At the same time, the surface mat outside the load is exposed to tensile stresses, which may result in the appearance of surface cracks parallel to the boundaries of the loaded area. This can lead ultimately to a punching failure, particularly if the peat is deposited on a weak substratum of soft clay or marl. Experience gained from the construction of highway embankments (Thurber 1965) indicates that the bearing capacity of muskeg does not generally increase when the thickness of the peat layer decreases, as would be the case if the soil failed in punching shear.

The stress distribution in the embankment fill causes the surface mat under a highway embankment to be exposed not only to vertical stress but also to large tensile stresses. If the tensile strength of the peat is not adequate, the surface mat breaks under the middle of the embankment and the fill material sinks into the soft decomposed peat or soft clay. The tensile stresses in the surface layer may be calculated by assuming that the fibrous surface peat acts as a continuous mat or beam on an elastic subgrade (Wyman 1950; Nevel 1961; Meyerhof 1960, 1962). The tensile strength of the surface layer can be determined either by direct tension tests or by beam tests (see Section 4.4(2)).

The ultimate punching pressure under a circular contact area on a homogeneous surface mat is approximately

$$p_u = (2 + m)m\sigma_t, \tag{18}$$

where σ_t is the tensile strength of the surface mat and $m = h/a$, where h is the thickness of the mat and a is the radius of the contact area. The local bearing capacity is approximately $3\sigma_t$ (Meyerhof 1960), which corresponds to the ultimate punching pressure in the case $m = 1$. It is also possible to calculate the critical load in the manner of Smith (1950) by using the theory of elasticity. If the ultimate load is assumed to be twice the critical load giving a maximum tensile stress equal to the tensile strength (σ_t) of the soil, the ultimate uniformly distributed load p_u on a circular contact area in the case $v = 0.5$ would be $4\sigma_t$. The p_u value increases when Poisson's Ratio v decreases; for $v = 0.4$, the ultimate load p_u would be about $4.5\,\sigma_t$ and for $v = 0.2$, it would be about $5.7\,\sigma_t$ (cf. Schiffman and Aggarwala 1960).

The factor of safety of an embankment on layered soil consisting of peat and clay layers is usually calculated by using a circular cylindrical surface of rupture. Because of the great difference between the structure and stress–strain properties of peat and clay, the shear resistance of these soils is not fully developed at the same time. The normal procedure of using peak strength values along the whole surface of rupture is, therefore, not justified in this case. It is possible for the ultimate resisting moment ΣM_u to be calculated from the peak value obtained from a $\Sigma s - \epsilon$ curve or be taken from a $\Sigma M - \epsilon$ curve, where Σs denotes the sum of the shearing resistance in the surface of rupture corresponding to a certain angular strain ϵ, and ΣM represents the resisting moment at a certain ϵ value.

B   Estimation of Bearing Capacity Based on Experience

Highway and railway embankments have been built across muskeg areas for many decades and much experience with the bearing capacity of muskeg has thus been obtained. Unfortunately, only a small part of this experience has been recorded and related to the strength properties of the peat. A few case records have been described, where the actual failure load was in reasonably good agreement with the ultimate bearing capacity calculated on the basis of unconfined compression tests or vane tests (Ward 1948; Anderson and Hemstock 1959; von Moos 1961).

In estimating the maximum height of fill that can be placed without producing failure, some engineers, on the basis of their experience, have used the formula

$$H_{ult} = 6\,\tau_f/\gamma \qquad\qquad (19)$$

(Lea and Brawner 1959; see also Section 6.5(4)), where τ_f is the shear strength of the peat measured by the vane and γ is the unit weight of the fill. Others (for example, Anderson and Hemstock 1959) have used the formula

$$H_{ult} = 5.5\,\tau_f/\gamma. \qquad\qquad (20)$$

Miyakawa (1959b) stated, on the other hand, that the normal interpretation of the circular arc method always results in an overestimation of the bearing capacity of the peat. Similar results were obtained by Ripley and Leonoff (1961), who reported that the actual bearing capacity of a deep peat deposit was only about 30–50 per cent of the theoretical value calculated on the basis of vane tests. With regard to the use of unconfined compression tests for estimating the bearing capacity of peat, Hanrahan (1964) expressed the opinion that these tests were somewhat unreliable for peat and tend to overestimate the strength of the material. Implications of the low K_0 value for peats in stability analyses and the development of shearing resistance on failure surfaces, have already been referred to in Section 4(1) of this chapter.

A purely empirical method for the estimation of the bearing capacity of peat has been proposed by Thurber (1965). On the basis of experience in highway construction in muskeg areas in Alberta and British Columbia, Thurber stated that the maximum permissible height of fill to avoid shear failure is 12 feet (3.66 m) if the depth of muskeg is less than 15 feet (4.58 m). If the depth of the muskeg (including both peat and soft clay) is greater than 15 feet (4.58 m), the maximum permissible height of fill would be 8 feet (2.44 m). This corresponds to a "safe bearing capacity" of about 1435 pounds per square foot (7 tons per sq m) and 1025 pounds per square foot (5 tons per sq m) respectively. As the strength and bearing capacity also depend on the type of muskeg and the underlying subsoil, however, this recommendation should be regarded as valid only for the range of muskeg types commonly occurring in Alberta and British Columbia (see Fig. 3.5).

There are, in fact, case records indicating smaller ultimate bearing capacities than the above recommended "safe" values. In a case in Hokkaido described by Miyakawa (1959b) failure occurred when the thickness of the fill was 2.4 m (6.56 feet). The case reported by Ripley and Leonoff (1961) is a further example of a smaller failure load. From Finland several cases are known (Helenelund 1953) where the actual failure load was smaller than the safe loads recommended above.

It is possible that a fairly good method for the estimation of the bearing capacity of peat could be developed on a purely empirical basis. Such a method should be based on observations in different types of muskeg so that the type of surface cover as well as the type and depth of peat and the underlying substratum can be fully considered. An empirical bearing capacity formula of the type

$$p_u = p_0 + cP/A \qquad\qquad (21)$$

has been suggested by Korchunov (1946), where p_0 for moss peat is about 2 to 4.5 tons per sq m (410 to 923 pounds per square foot) and c is a perimeter shear coefficient varying between about 0.5 and 0.9 tons per m (336 and 605 pounds per foot). P denotes the perimeter of the loaded area in metres and A is the area of the contact surface in square metres. For a rather long embankment with a width, for example, of 10 m (32.8 feet) or more, the influence of the perimeter shear is very small and the ultimate bearing capacity p_u is approximately equal to p_0. Both p_0 and c depend on the type of peat as well as on the degree of decomposition and the water content. These values, however, have been measured only for small loading plates and may, therefore, not be directly applicable to large-scale structures.

(5) *Consolidation and Settlement*

In this section the compressibility of peat is considered; the behaviour described also applies generally to organic soils when the organic component is the major constituent.

The major involvement of peat in engineering work is in its use as a foundation material. In this role the high compressibility of the material stands out as a most significant engineering property. In defining the compressibility of peat it is important to differentiate between deformations resulting purely from volume change and those resulting from lateral displacement (shear strain). The "consolidation" of peat will now be considered as the total compression resulting from volume change under a vertical load. The "settlement" will be considered as the vertical deformation of the muskeg surface. The vertical deformation resulting from displacement or shear may be appreciable, particularly at high load intensities, and these should be considered separately when predicting field compression from laboratory consolidation tests (see Section 4.4(3)).

One of the most striking differences in the compression of peat and organic soils as compared to mineral soils is the long-term compression which appears to be an almost continuous process. Probably the best examples of this long-term consolidation are some of the world famous Holland dykes, many of which are founded on peat (Buisman 1936). Records taken since 1880 have shown continuous settlement, showing for the most part a straight-line relation with the logarithm of time.

Because of the high compressibility and low strength of peat, its use as a foundation material has been limited to the support of low embankments, principally in road construction. In recent years, with the use of preloading techniques, however, the support of major highways, oil tanks, buildings, and even small earth dams has been possible on organic foundations.

A   Properties which Influence the Compressibility of Peat

The reasons for the high compressibility of peat are fairly obvious; however, a brief re-examination of some of the fundamental properties of the material will prove helpful in understanding its behaviour during consolidation.

As was pointed out in Section 4.1, peat is composed of organic solids, gas, and water. A small fraction of mineral soil, and in some instances a quite high fraction, may be present. In the latter case the material would be more accurately classified as an organic soil.

(i) TYPE   The classification of peat as given in Chapter 2 subdivides the material into three main divisions: amorphous-granular, fine-fibrous, and coarse-fibrous. A further breakdown of each major division according to structure is made, with a total of 17 categories (Table 2.1). Generally, the fine-fibrous peats will show the highest compressibility, the coarse-fibrous peats the least.

(ii) WATER CONTENT   The water content of peat is one of the most useful indicators of relative compressibility. The water content varies widely (see Table 4.2) and may range between 75 and 95 per cent by volume of water. Although the peat types represented by both extremes are highly compressible, the difference in compressibility will be clearly recognized.

(iii) PERMEABILITY   In the natural state most peats have a high porosity and are, therefore, pervious. For this reason the initial compression of peat occurs rapidly. It was shown in Section 4.2(3) that, as compression proceeds, the permeability is rapidly reduced. Even under moderate compressive loading the permeability change can be several orders of magnitude. The change in permeability during compression has, therefore, a profound effect on the rate of consolidation.

(iv) MINERAL SOIL CONTENT   The presence of mineral soil will temper the behaviour of peat depending on the amount present. A knowledge of the mineral soil content is instructive, therefore, in interpreting test information.

Other properties which will influence compressibility are gas content, specific gravity and density. These properties and those mentioned above are dealt with more thoroughly in Section 2 of this chapter.

B   Application of Conventional Soil Mechanics Consolidation Theory

For inorganic soils the basic assumptions made in applying the conventional consolidation theory are as follows: (i) homogeneous material, (ii) complete saturation, (iii) negligible compressibility of the solid matter, (iv) the validity of Darcy's Law, and (v) constant properties during each stage of consolidation; in particular, permeability.

In applying the conventional theory to the consolidation of peat, there are two major deviations from the above assumptions, namely, the compressibility of the solids and the change in permeability under applied load. These two anomalies are believed to account for the significant differences in consolidation behaviour between organic and mineral soils.

A typical laboratory time–settlement curve for peat is shown in Figure 4.7. It will be seen that an initial compression occurs over a very short time interval, followed by a long-term compression that is essentially straight with the logarithm of time. This curve is typical of many peats and it is generally found that the dissipation of porewater pressure is associated with the rapid or initial stage. For the most part, rapid dissipation of water pressure has also been found in the field. It may be inferred, therefore, that the initial stage is analogous to "primary consolidation" and the long-term stage to "secondary compression." The large magnitude and short duration of the initial stage and the continuous long-term compression are the major departures from mineral soil behaviour. The very high initial porosity is believed to account for the rapid and large increment of initial compression; the compressibility of the solid material is believed to account, at least in part, for the continuous long-term compression (Adams 1965).

Certainly with respect to the total compression of peat, the application of conventional consolidation theory does not appear warranted. It may be possible to apply the theory in a limited way, however, to the initial or primary stage.

c   Laboratory Analysis

(i) SAMPLING PROCEDURES   The obtaining of a reasonably "undisturbed" sample of peat is essential if laboratory testing is to be meaningful. The most satisfactory method of obtaining an undisturbed sample of peat is to cut a block or cylindrical sample in an open test pit. The disadvantage of this method is that it will be limited to a fairly shallow depth. Fair success has been claimed for the use of conventional thin-walled samplers with piston attachments. The Swedish foil sampler has been used to a limited extent. The inside diameter of the samplers used has normally been 2.5 to 3.0 inches (6.35 to 7.62 cm). Open drive thin-walled samplers have not been found satisfactory. Because of the inevitable compression which will occur for any drive sampling, it will be extremely important to record the recovery ratio. The development of a special sampler for recovering peat samples for undisturbed laboratory tests is clearly needed.

(ii) TESTING PROCEDURES   The consolidation testing of peat has commonly followed conventional practice with certain modifications. The sample diameter has varied from 2.5 to 12 inches (6.35 to 30.5 cm). The procedures which have been reported will be briefly discussed as follows:

(a) Conventional method, load increment ratio $\Delta p/p = 1$, load duration 24 hours. When this method is used, an appreciable magnitude of secondary (long-term) compression will occur under each load increment. It has been found for the most part that the rate of long-term compression of peat (slope of log-time plot) is proportional to the thickness. This was first observed by Buisman (1936). The magnitude of secondary compression under each individual increment will, therefore, have an effect on the rate of long-term compression on successive increments (Adams 1963). Consequently, for this method an allowance for the long-term compression component under each increment must be made.

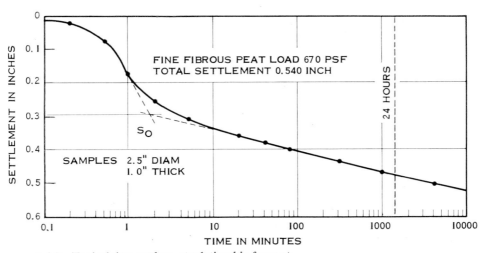

FIGURE 4.7   Typical time-settlement relationship for peat

(b) Single increment loading. In this method only one increment of load is applied to each specimen. The load duration may be 24 hours or more as desired. The results of this test are applied directly to field cases for similar loading intensity. The disadvantage of this method is that it requires a greater number of tests on individual specimens to cover the range of loading intensities required.

(c) Multiple increment loading of short duration. This method is used for the determination of rate and magnitude of the initial compression. Successive load increments are applied as soon as the pore water pressure dissipation period is complete. Loads are applied, therefore, at intervals varying from a few minutes up to about one hour, depending on the initial permeability and thickness of the peat specimen. This method will not provide information on the rate of long-term compression. Separate tests, of at least 24-hour duration, must be made to cover the load intensity under consideration.

(iii) RECOMMENDED TEST METHOD   The diameter of the sample tested should be as large as practicable. Because of the large compression that will occur, the thickness ratio should be greater than that used with mineral soils. The recommended ratio of diameter to thickness is 3. The recommended minimum sample diameter is 2.5 inches (6.35 cm).

In view of present knowledge, it is recommended that loads be applied as single increments covering the loading range desired. Measurement of settlement should be made over very short time intervals during the first few minutes of loading and then at conventional intervals. The load should be left on for a period of 24 hours. Multiple increment loading procedures may be possible, but more development work is required to correlate the sample thickness with the rate of long-term compression.

The results of individual tests should be plotted on charts of settlement versus the logarithm of time. The initial compression S_0 should be computed according to conventional practice, as shown in Figure 4.7. If a defined S curve is not obtained, the initial compression should be taken at the point where the settlement first becomes linear with the logarithm of time. The rate of long-term settlement should be expressed as C_s (coefficient of secondary compression). This is the slope of the settlement–log time plot divided by the thickness of the peat sample at the beginning of the long-term or straight-line stage. The initial compression may be expressed as a percentage of the initial thickness and plotted against the logarithm of the applied load. Similarly the coefficient of secondary compression may be plotted against the logarithm of applied load. Typical plots are shown in Figures 4.8 and 4.9.

(iv) INTERPRETATION OF RESULTS   For reasons already indicated, it has been found that the conventional curve-fitting methods of interpreting consolidation tests on peat or organic soils will not be possible. Several empirical approaches have been used with peat which have met with some degree of success. These will be mentioned briefly, indicating, where possible, their limitations and general application.

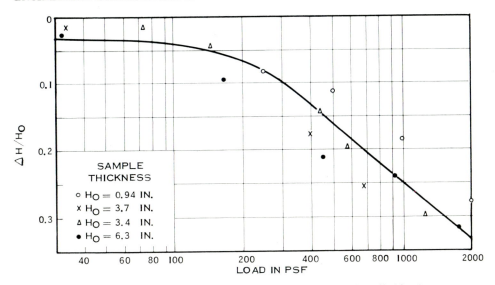

FIGURE 4.8   Initial settlement of peat in relation to peat thickness and applied load

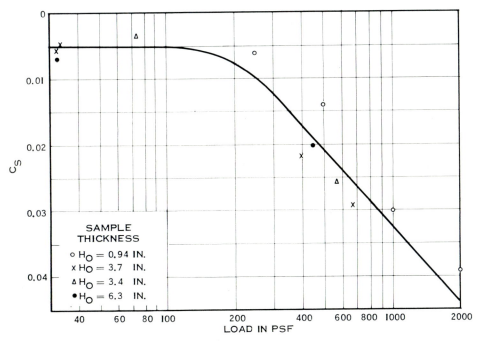

FIGURE 4.9   Rate of long-term consolidation of peat in relation to peat thickness and applied load

The magnitude of the initial settlement from laboratory tests usually has been found to be proportional to the thickness of the peat (Lea and Brawner 1963; and others). The duration of this phase has been found to be extremely short and in many field cases does not exceed the construction period. Lea and Brawner (1959) reasoned that this phase, being largely dependent on the drainage of water under excess hydrostatic pressure, should behave according to the classical Terzaghi theory and that the rate of dissipation would be a function of the square of the length of drainage path or peat thickness. In applying this reasoning to several embankments on peat in British Columbia, it was found that the time rate was proportional to the peat thickness to a power closer to 1.5 (Lea and Brawner 1963). This approach is followed in Chapter 6 of this Handbook in the discussion on preconsolidation.

For single increment tests, where the applied load has been made equal to the anticipated field loading, the magnitude and rate of the initial compression may be estimated as follows:

$$S_0 \text{ field} = \frac{H_0 \text{ field} \times S_0 \text{ lab}}{H_0 \text{ lab}}, \tag{22}$$

$$t \text{ field} = \frac{H_0{}^i \text{ field} \times t \text{ lab}}{H_0{}^i \text{ lab}}, \tag{23}$$

where t is the time for consolidation, H_0 the initial peat thickness, S_0 the initial compression, i the exponential parameter (generally 1.5, but may be as high as 2.0). Where a relation such as that given in Figure 4.8 has been developed, the magnitude of the initial compression may be computed for any load intensity and peat thickness.

The magnitude of the secondary or long-term compression, being continuous with time, is dependent on the time period under consideration. It has been found that the rate of long-term compression is proportional to the applied load and the peat thickness at the start of the long-term period (Adams 1964). The relation between laboratory and field rates is achieved by the use of the coefficient of secondary compression C_s. The total field settlement for any given time is then calculated as follows:

$$S_T = S_0 + C_s H \log_{10} t/t_0, \tag{24}$$

where S_T is the total settlement, S_0 the initial compression (primary), C_s the coefficient of secondary compression, t the field time considered, t_0 the estimated field time for initial compression, and H the thickness of peat layer at time t_0.

The prediction of field settlement from laboratory tests on peat is essentially a direct extrapolation of laboratory thickness to field thickness. It is based on the assumption of linear rate of settlement with the logarithm of time. This is not always the case, but field records as a rule have confirmed this behaviour for most practical purposes. There are many factors that can lead to erroneous predictions.

These include non-homogeneity of the peat mass, the effect of the underlying mineral soil layers, shear or displacement strains, and gas content. Considering these factors and the empirical nature of the relations used, one must realize that the predictions of field settlements will be extremely approximate. A further reference to predicting settlement from considerations of both primary consolidation and secondary compression is given in Section 7.1.

D   Practical Considerations

(i) PRELOADING   The most significant engineering characteristic of peat is that of continuous long-term compression. To overcome this difficulty, the use of preloading has been used successfully to eliminate or partially eliminate the long-term compression. A preload is applied, of sufficient magnitude and duration to cause the compression of the peat which would normally occur under the proposed design load over the expected life of the structure (see also Section 6.5(4)). Fortunately, peat lends itself ideally to preloading because of its high initial permeability. The required compression can usually be achieved, therefore, in a relatively short period of time. The magnitude of the preload will be limited by stability considerations which are largely governed by the rate of porewater pressure dissipation.

A method of estimating the magnitude and duration of the preload is shown in Figure 4.10 for a fine-fibrous peat 10 feet (3.05 m) in thickness. Curve No. 1 is the field time–settlement curve, computed according to the procedures suggested in

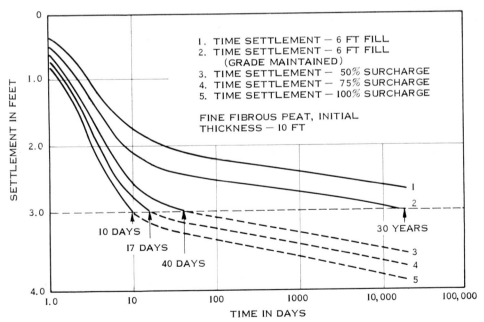

FIGURE 4.10   Method of estimating preload requirement using time-settlement curves

the preceding pages for a 6-foot (1.83-m) fill during a 30-year design life. Curve No. 2 is similar but allows additional fill to maintain grade. For this case, the amount of settlement calculated from curve No. 1 is considered as additional fill (submerged weight is used if the water table is at or close to the ground level). Curves 3, 4, and 5 are for embankment heights equivalent to 50, 75, and 100 per cent of the surcharge. The duration of application of these loads is taken where the 30-year settlement line from curve 2 intersects each of the preload curves.

For the case shown, a relatively pervious peat is considered. For more impervious peats a considerably longer period of surcharging would be required. The surcharge selected must be checked for stability against shear failure. A form of stage construction or the use of berms may be required if stability considerations will not allow the desired magnitude of surcharge loading. It should be realized that the surcharging of a peat foundation because of the relatively short duration will have little effect on the underlying soil, and appreciable long-term consolidation may still occur if compressible clay layers are present. Considerable field rebound has been observed in several field cases on removal of the preload. This appears to be most pronounced for loads reduced to a low final magnitude.

The preloading of peat foundations has been carried out infrequently and time has not allowed a complete assessment of the method. The greatest use of preloading has been in British Columbia (Lea and Brawner 1963). The effectiveness will depend on the ratio of surcharge to final load. This ratio will depend on the final design height, the thickness of the peat and nature of the underlying soil. A ratio of between 1.5 and 2.0 has frequently been used. The duration of the surcharge for peat thickness between 10 and 30 feet (3.05 and 9.15 m) has varied from a few weeks up to one year. For very deep peats the feasibility of preloading is quite uncertain.

(ii) SAND DRAINS   Sand drains have commonly been used in clay foundations to accelerate the initial or primary consolidation. They have been particularly useful in the preloading technique where the duration of the preload has considerable economic importance.

Opinion has been divided on the effectiveness of sand drains in peats and organic soils. In most cases, they have not been found to be particularly effective (Lake 1960; Lea and Brawner 1963). Sand drains in a foundation allow easier access for escaping water and correspondingly reduce excess water pressures in the foundation. As mentioned earlier, for many peats the initial or excess porewater pressure period is very short, occurring mainly within the construction period. It is obvious that for this condition sand drains will not be effective. For the long-term or secondary compression also the use of sand drains will not be effective since the rate of compression does not appear to be a function of length of the drainage path.

The use of sand drains should not be ruled out for all peats or organic soils, however. Where tests have shown that the duration of the initial or primary compression stage is appreciable and the length of drainage path (or thickness) is also large, sand drains could be effective both in reducing construction pore pressure and in accelerating the primary stage. The use of drains will have the further effect of stiffening a peat mass and improving the stability of the foundation.

(iii) SURFACE MAT   The strength and thickness of the surface mat of living vegetation has considerable influence on the stability of a peat foundation but little influence on the compressibility of the underlying peat. The thickness of the surface mat should not be considered in the long-term settlement prediction. The mat is usually coarse and will consolidate rapidly under load.

(iv) UNDERLYING SOIL   In many muskeg foundations, the underlying soils may be more dangerous from the standpoint of stability and settlement than the peat. If clays are present under the organic soils, they are frequently normally consolidated and, because of the negligible overburden pressure acting on them, may be in a very compressible or almost fluid condition. The problem of the foundation then may well lie in the stability and compressibility of these underlying soils. It is beyond the scope of this Handbook to deal with this condition, other than to point out that it is potentially dangerous and requires detailed study.

(v) INSTRUMENTATION   Knowledge of the compressibility of organic soils is limited. It is desirable, therefore, to measure, wherever possible, the pore water pressures and rate of consolidation in the field. These measurements can be made quite inexpensively and will invariably be of practical significance on the particular job as well as providing a valuable addition to the limited field of knowledge on the compressibility of organic materials. Examples of instrumentation are given in Sections 6.5(4) and 6.16.

4.5   CORRELATION OF ENGINEERING PROPERTIES OF PEAT

It would be helpful if various peat properties could be correlated with some easily determined characteristic such as water content, so that further detailed testing would not need to be carried out. Various investigators have attempted to make such correlations since some simple relations do exist. These relations permit a preliminary estimate to be made of settlement or shear, but in no way do they take the place of a careful field and laboratory investigation. The relations shown in the various correlation curves represent data from a wide variety of peat types as well as from some "organic soils" (mineral soils with the inclusion of a large amount of organic material). For the sake of clarity, the graphs generally do not show the data points, but only the smooth curve through these points. For more detailed and accurate information, therefore, the reader is directed to the references cited.

(1)   *Specific Gravity and Water Content*
Figure 4.11 shows that in the pure peat range (specific gravity averaging about 1.6) there is little relation with water content. With higher specific gravities ranging towards inorganic soils, however, there is a linear relation with water content, the specific gravity increasing as the water content decreases. This relation holds only

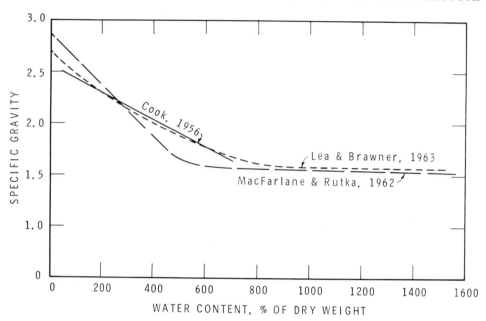

FIGURE 4.11.   Specific gravity vs. water content

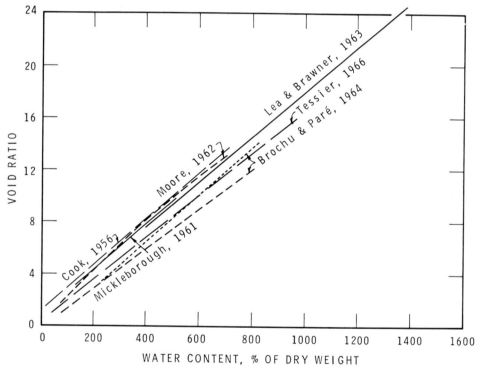

FIGURE 4.12   Void ratio vs. water content

for normally consolidated virgin peats. If peat is dried or compressed, the relation changes drastically (Lea and Brawner 1963).

(2)   *Void Ratio and Water Content*
The linear relation of Figure 4.12 can be anticipated for water content and void ratio since it is a simple mathematical relation involving specific gravity and gas content.

(3)   *Density and Water Content*
Curves of dry density, wet density, and submerged density from data supplied by three different investigators are shown on a log–log plot in Figure 4.13. The graph for dry density versus water content, however, represents surface layers of peat only to a depth of 2 feet (0.61 m).

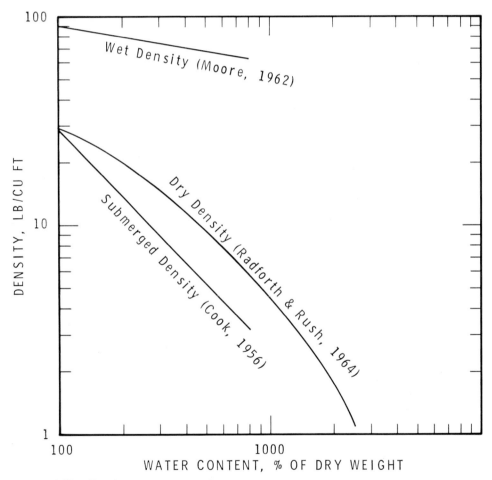

FIGURE 4.13.   Density vs. water content

(4) *Organic Content and Water Content*

In the pure peat range (averaging about 90 per cent organic content), there is little relation between organic content and water content. For lower organic contents, however, there is an apparent linear relation, as indicated in Figure 4.14, with the water content decreasing as the organic content decreases.

(5) *Organic Content and Acidity*

Although there is a considerable scatter of data points, there is some indication of a trend towards higher acidity with increase in organic content in the peats with an organic content of less than about 80 per cent. In the pure peat range of organic contents, however, there is little evidence of a relation between the two properties (Fig. 4.15).

(6) *Void Ratio and Permeability*

As was indicated in Section 4.2(3), the permeability of peat is greatly reduced with increased load. Lea and Brawner (1963) report that the permeability coefficient of virgin BC peat generally ranges from 10^{-2} to 10^{-4} cm per second. Under a load equivalent to only a few feet of fill, the permeability coefficient reduces to about 10^{-6} cm per second and under loads equivalent to 6 to 8 feet (1.83 to 2.44 m) of fill, k is reduced to 10^{-8}–10^{-9} cm per second. Their investigations resulted in the log–log plot of Figure 4.16 which indicates a straight-line relation between void ratio and coefficient of permeability.

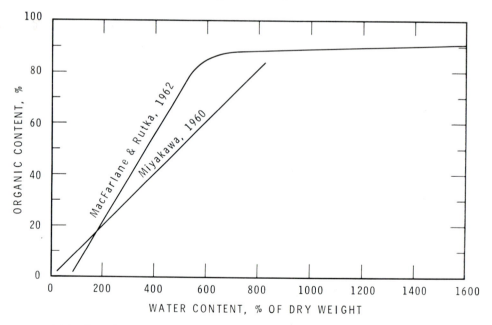

FIGURE 4.14   Organic content vs. water content

(7)   *Shear Strength and Water Content*

The shear strength – water content relation is reflected by the data in Figure 4.17 (from Anderson and Hemstock 1959). Despite a wide scatter of the test results, this figure shows an increase in shear strength with decreasing water content. In the discussion on the shear vane in Section 5.2(1), however, it is pointed out that this relation does not always obtain.

The range for the strength of remoulded peat is seen to be largely independent of the water content. The loss of strength due to remoulding is much greater for the lower range of water contents; the remoulded strength drops to about one-third of the undisturbed strength. In the highest water content range, the loss in strength due to remoulding is about 50 per cent.

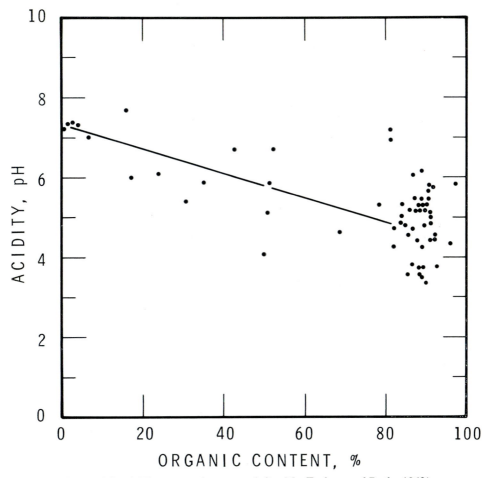

FIGURE 4.15   Acidity (pH) vs. organic content (after MacFarlane and Rutka 1962)

(8)   *Compression Index and Water Content*
Variations in the numerous factors which influence the consolidation characteristics of peat (see Section 4.4(5)) lead to some scatter of data in Figure 4.18. The relation between the compression index (C_c) and water content varies between $C_c = 0.0075\ w$ and $C_c = 0.011\ w$, with the average being about $C_c = 0.01\ w$.

(9)   *Compression Index and Void Ratio*
The observed relation between void ratio and consolidation properties is shown in Figures 4.19 and 4.20. These relations can be used for preliminary estimates in those areas represented by the references cited, but it is evident that the relation is somewhat different for the different localities. It is suggested (Cook 1967) that in any area at least two or three laboratory tests be carried out to establish the general relation.

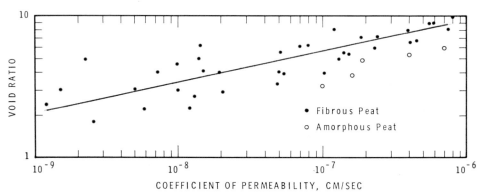

FIGURE 4.16   Void ratio (e) vs. coefficient of permeability (k) (after Lea and Brawner 1963)

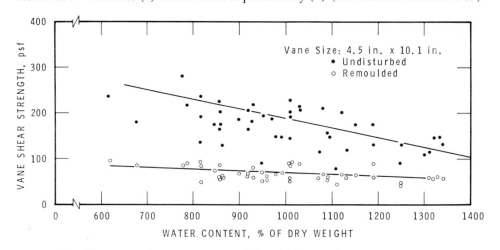

FIGURE 4.17   Shear strength vs. water content (after Anderson and Hemstock 1959)

It is evident that some relation should exist between the compression index (C_c) and void ratio. Figure 4.19 indicates a range for the various locations between $C_c = 0.45\ e$ and $C_c = 0.75\ e$.

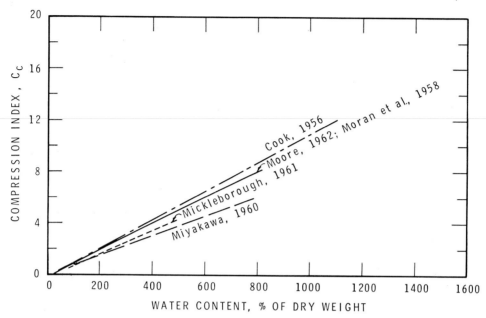

FIGURE 4.18   Compression index (C_c) vs. water content

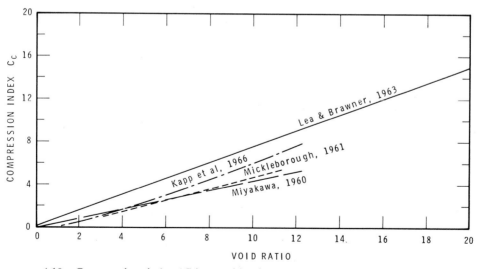

FIGURE 4.19   Compressions index (C_c) vs. void ratio

(10)   *Coefficient of Compressibility and Void Ratio*

In Figure 4.20 a non-linear relation is shown to exist between the coefficient of compressibility (a_v) and the void ratio.

A great amount of information is now available on the engineering characteristics of peat, but it is evident that many gaps in the knowledge still exist and that much of the available information is contradictory or confusing. In certain aspects of the muskeg problem, specific detail is virtually non-existent, for example, geophysical properties of peat. Correlation of physical properties with peat type is still in its infancy. Although extensive information is available on strength and deformation properties of peats, in many cases design is still somewhat empirical. Consequently, there is ample scope for additional research on both the fundamental and applied aspects of the physical properties of peat.

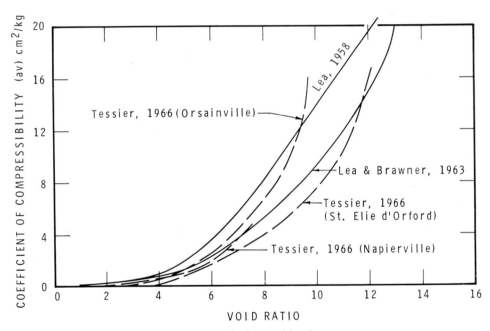

FIGURE 4.20   Coefficient of compressibility (a_v) vs. void ratio

REFERENCES

ADAMS, J. I. 1961. Laboratory compression tests on peat. Proc. Seventh Muskeg Res. Conf., NRC, ACSSM Tech. Memo. 71, pp. 36–54.

——— 1963. The consolidation of peat – field and laboratory measurements. Ontario Hydro Res. Quart., Vol. 15, Fourth Quarter, pp. 1–7.

——— 1964. A comparison of field and laboratory consolidation measurements in peat. Proc. Ninth Muskeg Res. Conf., NRC, ACSSM Tech. Memo. 81, pp. 117–135.

———— 1965. The engineering behaviour of a Canadian muskeg. Proc. Sixth Intern. Conf. on Soil Mech. and Foundation Eng., Vol. 1, pp. 3–7.

AKROYD, T. N. W. 1957. Laboratory testing in soil engineering. Soil Mechanics Limited, London, 233 pp.

ANDERSON, K. O. and R. C. HAAS. 1962. Construction over muskeg on the Red Deer Bypass. Proc. Eighth Muskeg Res. Conf., NRC, ACSSM Tech. Memo. 74, pp. 31–41.

ANDERSON, K. O. and R. A. HEMSTOCK. 1959. Relating some engineering properties of muskeg to some problems of field construction. Proc. Fifth Muskeg Res. Conf., NRC, ACSSM Tech. Memo. 61, pp. 16–25.

ASSOCIATE COMMITTEE ON SOIL AND SNOW MECHANICS. 1954–1962. Proc. Annual Can. Muskeg Res. Conf., NRC, ACSSM Tech. Memos 38, 42, 47, 54, 61, 67, 71, 74, Ottawa.

BOELTER, D. H. 1965. Hydraulic conductivity of peats. Soil Sci., Vol. 100, No. 4, pp. 227–231.

BROCHU, P. A. and J. J. PARÉ. 1964. Construction de routes sur tourbières dans la Province de Québec. Proc. Ninth Muskeg Res. Conf., NRC, ACSSM Tech. Memo. 81, pp. 74–108.

BROWN, R. J. E. 1963. Influence of vegetation on permafrost. Proc. Permafrost Intern. Conf., Lafayette, Indiana. Nat. Acad. Sci.-NRC Publ. No. 1287, Washington, DC, pp. 20–25.

BUISMAN, A. S. K. 1936. Results of long duration settlement tests. Proc. First Intern. Conf. on Soil Mech. and Foundation Eng., Vol. 1, pp. 103–106.

BURMISTER, D. M. 1959. Strain-rate behavior of clay and organic soils. Am. Soc. Testing Materials, Spec. Tech. Publ. No. 254, pp. 88–105.

CASAGRANDE, L. 1966. Construction of embankments across peaty soils. J. Boston Soc. Civil Engrs., Vol. 53, No. 3, pp. 272–317.

COLLEY, B. E. 1950. Construction of highways over peat and muck areas. Am. Highways, Vol. 29, No. 1, pp. 3–6.

COOK, P. M. 1956. Consolidation characteristics of organic soils. Proc. Ninth Can. Soil Mech. Conf., NRC, ACSSM Tech. Memo. 41, pp. 82–87.

———— 1967. Personal communication.

DE VRIES, D. A. 1963. Thermal properties of soils. In Physics of Plant Environment, edited by W. R. Van Wijk, chap. 7. John Wiley and Sons, New York, 382 pp.

DOYLE, R. G. 1963. The consolidation characteristics of peat. Unpublished MASc thesis, Dept. of Civil Eng., University of British Columbia. 110 leaves.

FARNHAM, R. S. 1957. The peat soils of Minnesota. Minnesota Farm and Home Science, Vol. XIV, No. 2, pp. 12–14.

FEUSTEL, I. C. and H. G. BYERS. 1930. The physical and chemical characteristics of certain American peat profiles. U.S. Dept. Agri., Bur. of Chem. and Soils, Tech. Bull. 214, Washington, DC.

GOODMAN, L. J. and C. N. LEE. 1962. Laboratory and field data on engineering characteristics of some peat soils. Proc. Eighth Muskeg Res. Conf., NRC, ACSSM Tech. Memo. 74, pp. 107–129.

GRAY, H. 1936. Stress distribution in elastic solids. Proc. First Intern. Conf. on Soil Mech. and Foundation Eng., Vol. 2, pp. 157–168.

HAEFELI, R. 1944. Erdbaumechanische Probleme im Lichte der Schneeforschung. Mitteilungen aus der Versuchsanstalt für Wasserbau an der ETH, No. 7, Zurich.

HANRAHAN, E. T. 1952. The mechanical properties of peat with special reference to road construction. Bull. Inst. Civil Engrs. Ireland, Vol. 78, No. 5, pp. 179–215.

———— 1954. An investigation of some physical properties of peat. Géotechnique, Vol. 4, No. 3, pp. 108–123.

———— 1964. A road failure on peat. Géotechnique, Vol. 14, No. 3, pp. 185–202.

HARDY, R. M. and S. THOMSON. 1956. Measurement of the shearing strength of muskeg.

Proc. Eastern Muskeg Res. Meeting, NRC, ACSSM Tech. Memo. 42, pp. 16–24.

HELENELUND, K. V. 1953. Stability and failure of the subsoil with special reference to railway embankments in Finland. State Inst. Tech. Res., Publ. No. 24. Helsinki.

———— 1967. Vane tests and tension tests on fibrous peat. Proc. Conf. on Shear Strength Properties of Natural Soils and Rocks, Vol. 1, Oslo, pp. 199–203.

HILLIS, C. F. and C. O. BRAWNER. 1961. The compressibility of peat with reference to major highway construction in British Columbia. Proc. Seventh Muskeg Res. Conf., NRC, ACSSM Tech. Memo. 71, pp. 204–227.

JACKSON, M. 1958. Soil chemical analysis. Prentice-Hall, Englewood Cliffs, NJ, 498 pp.

JUMIKIS, A. R. 1962. Soil Mechanics. D. Van Nostrand Company, Inc., Princeton, NJ, 791 pp.

JURGENSON, L. 1934. The application of theories of elasticity and plasticity to foundation problems. J. Boston Soc. Civil Engrs. See also Boston Soc. Civil Engrs. Contributions to Soil Mechanics, 1925–1940, pp. 148–183.

KAPP, M. S., D. L. YORK, A. ARONOWITZ, and H. SITOMER. 1966. Construction on marshland deposits: treatment and results. Highway Res. Board, Highway Res. Rec. No. 133, Washington, DC, pp. 1–22.

KERSTEN, M. S. 1959. Frost penetration: relationship to air temperature and other factors. Highway Research Board, Bull. No. 225, Washington, DC.

KORCHUNOV. 1946. Bearing capacity of deposits. Torfyanaya Promyshlennost, No. 9, p. 22. Bord na Mona E. S. Transl., No. 4, Dublin.

LAKE, J. R. 1960. Pore pressure and settlement measurements during small-scale and laboratory experiments to determine the effectiveness of vertical sand drains in peat. Proc. Conf. on Pore Pressure and Suction in Soils, Butterworth, London, pp. 103–107.

———— 1961. Investigation of the problem of constructing roads on peat in Scotland. Proc. Seventh Muskeg Res. Conf., NRC, ACSSM Tech. Memo. 71, pp. 133–148.

LAMBE, T. W. 1964. Methods of estimating settlement. Proc. Amer. Soc. Civil Engrs., J. Soil Mech. and Foundations Div., Vol. 90, No. SM5, pp. 43–67.

LEA, F. M. 1956. The chemistry of cement and concrete (rev. ed. of Lea and Desch). Edward Arnold (Publishers) Ltd., London, 637 pp.

LEA, N. D. 1958. Notes on the mechanical properties of peat. Proc. Fourth Muskeg Res. Conf., NRC, ACSSM Tech. Memo. 54, pp. 53–57.

———— 1962. Recent experience in major highway and construction developments over peat lands near Vancouver, B.C. Proc. Eighth Muskeg Res. Conf., NRC, ACSSM Tech. Memo. 74, pp. 59–67.

LEA, N. D. and C. O. BRAWNER. 1959. Foundation and pavement design for highways on peat. Proc. Fortieth Convention, Can. Good Roads Assoc., Ottawa, pp. 406–424.

———— 1963. Highway design and construction over peat deposits in lower British Columbia. Highway Res. Board, Res. Rec. No. 7, Washington, DC, pp. 1–33.

LEWIS, W. A. 1956. The settlement of the approach embankments to a new road bridge at Lockford, West Suffolk. Géotechnique, Vol. 6, No. 3, pp. 106–114.

LO, M. B. and N. E. WILSON. 1965. Migration of pore water during consolidation of peat. Proc. Tenth Muskeg Res. Conf., NRC, ACSSM Tech. Memo. 85, pp. 131–142.

LYTTLE, S. A. and B. N. DRISKELL. 1954. Physical and chemical characteristics of the peats, mucks and clays of the coastal marsh area of St. Mary Parish, Louisiana. Louisiana Agri. Exper. Sta., Bull. No. 484.

MACFARLANE, I. C. 1962. The engineering characteristics of peat. Proc. Atlantic Prov. Reg. Seminars on Organic Terrain Problems, NRC, ACSSM Tech. Memo. 77, pp. 45–50.

———— 1966. Peat structure as a basis of classification. Proc. Eleventh Muskeg Res. Conf., NRC, ACGR Tech. Memo. 87, pp. 26–31.

MACFARLANE, I. C. and C. M. ALLEN. 1964. An examination of some index test procedures for peat: a progress report. Proc. Ninth Muskeg Res. Conf., NRC, ACSSM Tech. Memo. 81, pp. 171–183.

MACFARLANE, I. C. and A. RUTKA. 1962. An evaluation of pavement performance over muskeg in northern Ontario. Highway Res. Board, Res. Bull. 316, Washington, DC, pp. 32–43.

MEYERHOF, G. G. 1960. Bearing capacity of floating ice sheets. Proc. Amer. Soc. Civ. Engrs., J. Eng. Mech. Div., Vol. 86, No. EM5, pp. 113–145.

———— 1962. Load carrying capacity of concrete pavements. Proc. Am. Soc. Civ. Engrs., J. Soil Mech. and Foundations Div., Vol. 88, No. SM3, pt. 1, pp. 89–116.

MICKLEBOROUGH, B. W. 1961. Embankment construction in muskeg at Prince Albert. Proc. Seventh Muskeg Res. Conf., NRC, ACSSM Tech. Memo. 71, pp. 164–185.

MIYAKAWA, I. 1959a. Soil engineering research on peaty alluvia: reports 1 to 3. Civil Eng. Res. Inst., Hokkaido Development Bureau, Bull. No. 20, Sapporo, 88 pp. English transl. by K. Shimizu, NRC Transl. no. 1001, Ottawa, 1962.

———— 1959b. On the stability of soft peaty ground under an earth embankment. Civil Eng. Res. Inst., Hokkaido Development Bureau, Rept. No. 21, Sapporo, 22 pp.

———— 1960. Some aspects of road construction in peaty or marshy areas in Hokkaido, with particular reference to filling methods. Civil Eng. Res. Inst., Hokkaido Development Bureau, Sapporo, 54 pp.

MOORE, L. H. 1962. A correlation of the engineering characteristics of organic soils in New York State (preliminary). New York State Dept. Public Works, Bureau of Soil Mech., Tech. Rept. 13 leaves.

MORAN, PROCTOR, MUESER, AND RUTLEDGE. 1958. Study of deep soil stabilization by vertical sand drains. US Dept. Navy, Bureau of Yards and Docks, Rept. no. y88812, Washington, DC, 429 pp.

MULLER, S. W. 1945. Permafrost or permanently frozen ground and related engineering problems. Office of Chief of Engrs., US Army, Spec. Rept., Strategic Engineering Study No. 62, Washington, DC, 231 pp.

NEVEL, D.E. 1961. The narrow free infinite wedge on an elastic foundation. Cold Regions Res. and Eng. Lab. Res. Rept. No. 79, Hanover, NH, 11 pp.

NEW SCIENTIST. 1964. Heat insulating sheet made from peat. vol. 22, 28 May, pp. 393–554.

OHIRA, Y. 1962. Some engineering researches on the experiments of the physical properties of the peat and on the sounding explorations of the peaty area in Hokkaido, Japan. Memoirs of the Defence Academy, Vol. II, No. 2, pp. 253–282.

PHALEN, T. E. 1961. Recent investigation concerning a high void ratio soil. Discussion in Proc. Fifth Intern. Conf. on Soil Mech. and Foundation Eng., Vol. 3, pp. 379–380.

PRIESTLEY, C. H. B. 1959. Turbulent transfer in the lower atmosphere. University of Chicago Press, Chicago, 130 pp.

RADFORTH, N. W. 1955. Personal communication.

RADFORTH, N. W. and E. S. RUSH. 1964. Trafficability tests on confined organic terrain (muskeg): Report 1: Summer 1961 tests. US Army Corps of Engrs., Waterways Exp. Sta., Tech. Rept. no. 3–656, Vicksburg, Miss., 110 pp.

RIPLEY, C. F. and C. E. LEONOFF. 1961. Embankment settlement behaviour on deep peat. Proc. Seventh Muskeg Res. Conf., NRC, ACSSM Tech. Memo. 71, pp. 185–204.

RISI, J. et al. 1950–55. A chemical study of the peats of Quebec. Québec Dept. Mines, Lab. Branch, 10 parts in 5 reports: PR 234, 281, 282, 301, and 306.

ROOT, A. W. 1958. California experience in construction of highways across marsh deposits. Highway Res. Board, Bull. 173, Washington, DC, pp. 46–64.

SCHIFFMAN, R. L. and B. D. AGGARWALA. 1960. Stresses and displacements produced in an elastic half-space by an elliptic punch. Rensselaer Polytech. Inst., Winslow Labs., Troy, NY, 45 pp.

SCHROEDER, J. and N. E. WILSON. 1962. The analysis of secondary consolidation of peat. Proc. Eighth Muskeg Res. Conf., NRC, ACSSM Tech. Memo. 74, pp. 130–144.

SHEA, P. H. 1955. Unusual foundation conditions in the Everglades. Trans. Amer. Soc. Civ. Engrs., Vol. 120, pp. 92–102.

SKAVEN-HAUG, S. V. 1959. Protection against frost heaving on the Norwegian railways. Géotechnique, Vol. 9, No. 3, pp. 87–106.

SMITH, A. H. V. 1950. A survey of some British peats and their strength characteristics. Army Operational Res. Group, Rept. No. 32–49, London.

SKEMPTON, A. W. and L. BJERRUM. 1957. A contribution to the settlement analysis of foundations on clay. Géotechnique, Vol. 7, No. 4, pp. 168–178.

STEINBRENNER, W. 1934. Tafeln zur Setzungsberechnung. Die Strasse, Vol. 1, pp. 121–124. See also Proc. First Intern. Conf. on Soil Mech. and Foundation Eng. Vol. 2, pp. 142–143, 1936.

TESSIER, G. 1966. Deux exemples – types de construction de routes sur muskegs au Québec. Proc. Eleventh Muskeg Res. Conf., NRC, ACGR Tech. Memo. 87, pp. 92–141.

THOMPSON, J. B. and L. A. PALMER. 1951. Report of consolidation tests with peat. Symposium on Consolidation Testing of Soils, ASTM Spec. Tech. Publ. No. 126, pp. 4–8.

THURBER, R. C. 1965. Some methods of construction in peat bogs and swamps in British Columbia. Proc. Tenth Muskeg Res. Conf., NRC, ACSSM Tech. Memo. 85, pp. 59–69.

TRESSIDER, J. O. 1958. A review of existing methods of road construction over peat. DSIR, RRL, Tech. Paper No. 40, London.

TVEITEN, A. A. 1956. Applicability of peat as an impervious material for earth dams (Norwegian, with English Summary). Norwegian Geotech. Inst., Publ. No. 14.

VAN MIERLO, W. C. and H. I. DEN BREEJE. 1948. Determination of expected settlements of hydraulic fills in the Spangen Polder near Rotterdam. Proc. Second Intern. Conf. on Soil Mech. and Foundation Eng., Vol. 2, pp. 125–130.

VON MOOS, A. 1961. Rutschungen eines Strassendammes in einem Torfgebiet bei Sargans, Kanton St. Gallen. Strasse und Verkehr., No. 1, p. 5.

WAKSMAN, S. A. 1942. The peats of New Jersey and their utilization. NJ Dept. Conservation and Development, Bull. 55, Geol. Series, Trenton, NJ, 155 pp.

WARD, W. H. 1948. A slip in a flood defence bank constructed on a peat bog. Proc. Second Intern. Conf. on Soil Mech. and Foundation Eng., Vol. 2, pp. 19–23.

WARD, W. H., A. PENMAN, and R. E. GIBSON. 1955. Stability of a bank on a thin peat layer. Géotechnique, Vol. 5, No. 2, pp. 154–163.

WILLIAMS, G. P. 1966. Soil freezing and thawing at the Muck Experiment Station, Bradford, Ontario. School of Agric. Eng., University of Guelph, Eng. Tech. Publ. No. 14.

WYLD, R. C. 1956. A further investigation of the engineering properties of muskeg. Unpublished M.Sc. thesis, Faculty of Engineering, University of Alberta, 76 leaves.

WYMAN, M. 1950. Deflections of an infinite plate. Can. J. Res., Vol. 28, Sect. A, pp. 293–302.

5 Preliminary Engineering Investigations

Chapter Co-ordinator *
J. R. RADFORTH

5.1 TERRAIN ANALYSIS

Since the design of engineering structures and the planning of off-road transportation are dictated by the terrain on which these activities take place, it is important to find out how and to what extent the terrain exerts its influence. The relative importance of various terrain characteristics varies with the particular engineering operation under consideration. In general, it can be said that muskeg poses potential problems over a wide area in Canada, as indicated by the map in Figure 3.5 (Radforth 1961).

The type of work involved in most terrain analysis can be briefly outlined as follows: (1) surface analysis – airphoto interpretation (see Chapter 3), ground level analysis; (2) subsurface analysis – sounding for depth, sampling to ascertain peat profile, peat type, etc., sampling to obtain laboratory test samples, peat property measurements.

The objectives of any terrain-analysis study depend on the specific task involving contact with the terrain. For example, terrain analysis for vehicle mobility studies is quite different from that for foundation studies. Table 5.1 lists individual types of operations as well as terrain features which affect these muskeg operations. Those terrain features which affect a particular operation are indicated in the table by an x placed opposite the operation in question. In the analysis for a particular operation, the table can be checked for the controlling features of the terrain which can then be examined for the presence or absence and frequency of occurrence of these features. When these factors have been established, the best approach to the accomplishment of the operation can be developed from the following chapters.

In the assessment of some terrain features, an indirect approach must be used, whereas in others the analysis may be carried out by direct observation. Table 5.2 lists the terrain features and corresponding methods by which individual features can be detected, observed, measured, or otherwise treated.

---

*Chapter Consultant: C. O. Brawner – Vane Test.

TABLE 5.1
Occurrence of influence of terrain features on engineering operations

	Construction					Off-road vehicles	
	Road	Railway	Pipeline	Building	Tower and pylon	Route selection	Design
Vegetal cover	X	X	X			X	X
Surface profile (microtopography)						X	X
Peat depth	X	X	X	X	X		X
Type of mineral sublayer	X	X		X	X		
Peat structure	X	X	X	X	X	X	X
Peat density	X	X	X	X	X		
Shear strength	X	X		X	X	X	X
Cone index						X	X
Peat water content	X	X	X	X	X	X	X
Peat temperature			X	X			
Drainage pattern	X	X		X		X	
Permeability	X	X		X			
Frequency of open water	X	X	X		X	X	X
Water acidity (pH)	X	X	X	X	X		
Permafrost	X	X	X	X	X	X	X
Seasonal ice						X	X

TABLE 5.2
Methods of observation and evaluation of terrain features

Terrain feature	Method of analysis
Vegetal cover	Direct observation (low altitude oblique airphotos), Radforth vegetal cover classification system
Surface profile (microtopography)	Surveyor's level, tape
Peat depth	Manual probing with rods, geophysical methods
Type of mineral sublayer	Soil sampler, auger
Peat structure	Visual inspection, von Post or Radforth classification system
Peat density	Nuclear density and moisture meter
Shear strength	Shear vane (gives estimate only)
Cone index	Cone penetrometer
Peat water content	Sampling and drying or nuclear moisture meter
Peat temperature	Thermistor or thermocouple probe
Drainage pattern	Airphoto interpretation
Permeability	Field – insert hollow tube in peat, remove peat and water from tube, time period required for water to rise in tube Lab. – measure time required for water under constant pressure to flow from top to bottom of peat sample
Frequency of open water	Airphoto interpretation
Water acidity (pH)	Sampling, pH meter
Permafrost	Airphoto interpretation, ground observation
Seasonal ice	Airphoto interpretation, ground observation

5.2   FIELD TESTING

Field testing can include measurements of permeability, density, water content, and tests with the cone penetrometer and shear vane as well as sampling with various types of sampler. The tests chosen will be determined by the particular problem and are done in order to establish the constraints placed on a particular project by the muskeg environment. The measurements and subsurface samples can often be augmented by visual observation of the topography and surface vegetation. The chart in Table 5.3 can be used to provide a qualitative indication of the probable terrain conditions through a knowledge of the vegetal cover.

(1)   *Vane Tests*

The conventional method of obtaining undisturbed soil samples and performing unconfined compression or triaxial tests has met with limited success with peat because of the extreme difficulty of obtaining representative undisturbed samples, the difficulty in preparing samples for testing, and the high strain-to-failure characteristics of peat. As a result, the *in situ* vane shear test has frequently been used to evaluate shear strength.

Field vane tests normally are carried out with a device similar to that shown in Figure 5.1. The blades on the bottom of the instrument are inserted vertically into the ground and the shaft is rotated manually with a torque wrench at its upper end. Ideally, the torque readings and vane blade dimensions are used to calculate the shear strength of the peat from the formula

$$\tau_f = \frac{3T}{2\pi r^2 (2r + 3H)},\tag{1}$$

where τ_f is the shear strength, T the torque reading, r the radius of shear vane, and H the height of shear vane.

There has been mixed reaction to the accuracy of the vane test in peats. Some concern has been expressed that drainage occurs rapidly during the test and results in too high a shear value. Also, in fibrous peats much of the torque applied to the vane is absorbed by elastic deflection of the fibres surrounding the vane, and it has been found (Helenelund 1967) that these fibres recover some of their original orientation when the vane has rotated through an angle equal to the angle between two adjacent blades.

To confuse the picture in fibrous peats even further, it is probable that when the vane is inserted in the peat some of the fibres are cut by it, thus destroying some of the peat shear strength at the location where the measurement is being taken.

In view of the wide use of the field vane in determining the shear strength of peats, it is proposed to assess here the merits of this test. A brief review of several Canadian case histories is presented and the results are discussed. Data from 16

TABLE 5.3
Vegetal cover vs. mechanical properties (after Brawner 1957)

Cover type	Peat category	Topography	Depth of deposit (ft)	Permeability*	Water content*	Compressibility	Thickness of surface mat (ft)	Strength of surface mat*	Probable mineral soil base	Max. vert. ht. peat will stand unsupported (ft)	Probable ht. of fill peat will support (ft)†	Drainage characteristics*	Description
AH	9, 16	b	$\frac{1}{4}$-1	1	12	12	$\frac{1}{2}$	8	Sand	Up to 8	1	1	Trees over 15 ft; non-woody leathery to crisp mats
AHE	9, 13	j	1-3	2	11	11	1-2	3	Clean coarse gravel	Up to 10	1	2	Trees over 15 ft; non-woody crisp leathery mats; woody shrubs 0-2 ft high
AEH	9	i	2-5	3	10	10	2-3	1	Dirty coarse gravel	Up to 10	1	4	Trees over 15 ft; woody shrubs 0-2 ft high; non-woody crisp leathery mats
AE	9	i	Up to 7	4	9	9	$1\frac{1}{2}$	2	Clay silt with sand gravel	Up to 8	1	5	Trees over 15 ft; woody shrubs 0-2 ft high
BHE	15	i, j	Up to 15	5	7	7	Up to 5	5	Silt to sand	6-8	$\frac{3}{4}$	3	Trees, 5-15 ft high; crisp leathery mats; woody shrubs 0-2 ft high
HE	3	p, c, i	Up to 30	8	5	5	Up to 3	11	Sand to silt	5-6	$\frac{1}{2}$	8	Crisp leathery mats; woody shrubs 0-2 ft high

BEI	11	i, b	Up to 20	7	6	6	Up to 6	4	Clay to silt	5–6	$\frac{1}{2}$	7	Trees, 5–15 ft high; woody shrubs 0–2 ft; soft velvety mats
FI	2	a, p, k b	Up to 20	12	1	1	$\frac{1}{2}$	12	Clay	Will not stand	$\frac{1}{10}$	11	Non-woody grasslike mats—clumps or patches 0–2 ft high; non-woody soft velvety mats
EH	9	i o, m, h	Up to 30	3	4	4	Up to 8	10	Clay to silt	4	$\frac{1}{2}$	9	Woody shrubs 0–2 ft high; non-woody crisp leathery mats
BEH	9	h, i	Up to 15	6	7	7	Up to 6	6	Silt to sand	6–8	$\frac{3}{4}$	6	Trees, 5–15 ft high; woody shrubs 0–2 ft high; leathery crisp mats
DFI	10	m, l, a	Up to 10	11	3	3	1	7	Clay silt to sand	2–3	$\frac{1}{2}$	12	Woody shrubs or very dwarf trees 2–5 ft high; non-woody grasslike mats—clumps or patches 0–2 ft; soft velvety mats
EI	11	i, c, o	Up to 40	10	2	2	Up to 10	9	Clay	3	$\frac{1}{3}$	10	Woody shrubs 0–2 ft high; soft velvety mats

*Denotes relative values; 1, highest; 12, lowest.
†Denotes relative values with 1 maximum.

projects in five provinces are summarized in Table 5.4. Most of the projects have been associated with investigations to evaluate the application of preconsolidation for highway construction in muskeg (see Section 6.5(4)).

The vane shear test results were used to evaluate the stability of the fill construction. In 11 cases out of 12 reported, the fills were safely constructed. In one case (Anderson and Hemstock 1959) the peat was purposely loaded to failure. An unplanned failure occurred in only one case (Ripley and Leonoff 1961).

The types of peat in which the vane tests were performed were generally amorphous-granular or fine-fibrous. The depth of the peat deposits ranged from 6 to 35 feet (1.83 to 10.68 m). The mineral soil underlying the muskeg was silt or clay at nine of the projects, sand at four sites, and till at one location.

Vane shear test results ranged from a low of 30 pounds per square foot (146.5 kg per sq m) to a high of 1100 pounds per square foot (5370.6 kg per sq m). The most common range was 150 to 300 pounds per square foot (732.4 to 1464.7 kg per sq m). Many investigators found that the water content of the peat at comparative depths ranged from 100 to 2000 per cent. The higher values were obtained in the peat near Vancouver, BC. Peat with a water content less than 500 per cent

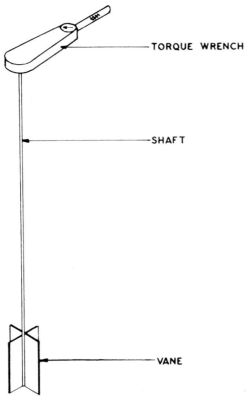

FIGURE 5.1   Vane tester for measuring shear strength of soils

TABLE 5.4
Summary of vane shear data from 16 projects in Canada

Reference	Location	Peat type	Peat depth (ft)	Base soil	Vane size (in.)	Shear rate (deg/sec)	Shear strength (p.s.f.)	Water content (%)	Sensitivity	Field fill failure	Applied effective pressure (p.s.f.)	Time to apply load (days)	Shear strength for max. load (p.s.f.)
Hardy and Thomson (1956)	North east BC	Fibrous	12–17	Clay	4 × 4.5	0.1	100–600	470–760	—	—	—	—	—
Brawner (1959)	Lulu Island, BC	Fibrous	6–10	Silt clay	3.5 × 7	1.0	150–260	1200–1700	1.1–2.5	No	650	15	120
Anderson and Hemstock (1959)	Near Edmonton, Alta.	Fibrous amorphous	10–14	—	4.5 × 10.1	0.5	100–220	625–1350	1.5–3.7	Yes	1220	—	220
MacFarlane (1960)	Northern Ont.	—	11	Clay	2 × 4 2.8 × 5.6 4 × 8	3 3 3	425–1100 350–600 200–300	675–1200 675–1200 675–1200	1.5–10	—	—	—	—
Ripley and Leonoff (1961)	Vancouver, BC	Amorphous fibrous	25	Sand	2.5 × 5.0	0.4	288–432	100–2000	3–4	Yes	700	60	140
Mickleborough (1961)	Prince Albert, Sask.	Fibrous	6	Sand	3.6 × 6	—	200–400	144–480	—	No	660	—	120
Anderson and Haas (1962)	Red Deer, Alta.	Fibrous amorphous	11	Clay	4.5 × 10.1	0.5	225–470	100–475	3–6	No	1990	110	345
Brochu and Paré (1964)	Orsainville, Napierville, Que.	Fibrous to amorphous	14 11	— Clay	3.0 × 6.0 3.0 × 6.0	12 12	100–550 125–650	150–700 100–600	4 3.5–4.5	No No	840 840	50 60	155 155
Lea and Brawner (1963)	Maillardville, BC	Fibrous	22	Clay silt	2.5 × 5.0	0.1	40–450	560–1200	—	No	930	21	170
Lea and Brawner (1963)	Deer Lake, BC	Fibrous amorphous	35	Clay	2.5 × 5.0	0.1	100–320	200–1200	—	No	500	40	90
Adams (1963)	Mattagami, Ont.	Fibrous	8–14	Till	2.0 × 3.0	—	200–800	200–600	—	No	500	120	90
MacFarlane (1964)	Ottawa, Ont.	Fibrous	7	Sand	2.8 × 5.6	—	150–370	—	—	—	—	—	—
Tessier (1966)	St. Elie d'Orford, Que.	Fibrous amorphous	13	Sand	2.6 × 6.0	12	120–420	200–890	—	No	800	72	145
Tessier (1967)	St. Bernard, Co. St. Jean, Que.	Fibrous to amorphous	11	Clay	2.2 × 4.4	—	190–420	—	—	No	—	—	—
	Napierville, Que.	Fibrous amorphous	11	Clay	3.0 × 6.0 2.2 × 4.4 3.0 × 6.0	12 12	30–315 170–500 100–390	100–600	—	No	1040 810	200 60	190 145

is believed to contain some mineral soil. Some investigators performed remoulded vane shear tests and found a sensitivity ranging from 1.1 to 10 with the most common values ranging between 1.5 and 4.0.

The average shear stress, τ_{av}, induced in the peat for all fills placed was computed by using an approximate bearing capacity formula of

$$\tau_{av} = p/5.5, \qquad (2)$$

where p is the effective load applied on the peat (pounds per square foot).

An accurate comparative analysis is not possible from the data, since the exact dimensions and profile were not given for most of the cases reported. In the 13 cases for which data are available, the applied average shear stress was below the lowest field shear test value obtained on five projects between the low and average shear values on six, and above the average on two. Successful use of the vane shear test for stability evaluation in 11 out of 12 projects suggests that the method has some merit. The one failure reported, however, indicates the need for review of the test method.

The main factors which are believed to influence vane shear test results are peat type, water content, and the test procedure. For practical engineering purposes, the Radforth peat categories (Chapter 2) can be reduced to three general classes, amorphous-granular, non-woody fibrous, and woody fibrous. Vane shear tests have shown reasonable consistency in the first two types but results in the latter are erratic. No numerical strength comparisons at the same site and with the same equipment were cited in the literature between amorphous-granular and fibrous peat. Evaluation of the data at the same locations, however, suggests that the amorphous-granular peat generally has lower shear strength (as indicated in Table 4.1).

Several investigators have reported that the vane shear strength varies with water content. Anderson and Hemstock (1959), for example, report a general decrease in strength from about 225 pounds per square foot (1099 kg per sq m) at 700 per cent moisture content to 110 pounds per square foot (537 kg per sq m) at 1400 per cent moisture content (see Fig. 4.17). The peat was amorphous-granular and fine fibrous in texture. A review of MacFarlane's data (1960) also indicates that the shear strength decreased with increased water content when a 2-inch (5.08-cm) diameter vane was used. On the other hand, he reports that the shear strength decreased with a decrease in water content when a 4-in. (10.16-cm) diameter vane was used (Fig. 5.2). The existing data on a possible relation between vane shear strength and water content are, therefore, somewhat inconclusive. Further research is clearly needed on this aspect.

MacFarlane (1960) and Tessier (1967) performed shear tests with various sizes of vanes. MacFarlane used vanes with diameters of 2.0 inches (5.08 cm), 2.8 inches (7.11 cm), and 4.0 inches (10.16 cm) (Fig. 5.2). He reports that there was a marked variation in results which was consistent for all the sites investigated.

The three curves (of vane shear plotted against depth) all have the same shape and clearly reflect any layers of higher or lower strength, but the smallest vane, on the average, gives results about double those of the medium-sized vane, and from four to five times those of the larger vane.

Tessier performed tests at three sites. Figure 5.3 shows comparative curves for a peat deposit 11 feet (3.34 m) deep at St. Bernard Co., St. Jean, Que. On the average, the 2.2-inch (5.59-cm) vane gave shear values 1.20 times those of the 2.6-inch (6.61-cm) vane and 1.44 times those of the 3.0-inch (7.62-cm) vane.

These results suggest that vane size is significant for tests in peat. The experience in clay is quite different. Cadling and Odenstad (1950) performed tests in Sweden and concluded that vane dimensions apparently do not affect the shear results for clay.

In the one case in Canada where the use of vane shear results led to a failure (Ripley and Leonoff 1961), it appears that the small vane that was used may have overestimated the shear strength of the peat. The lower shear value of 288 pounds per square foot (1407 kg per sq m) reported is considerably higher than the lower range of values of 40 to 150 pounds per square foot (195.3 to 732.4 kg per sq m) obtained by others in the same general area in British Columbia.

In clay, the normal rate of strain used in the shear test is 0.1 degrees per second. This rate was used on three of the projects cited. On 10 of the projects more rapid test rates, ranging up to 12 degrees per second, were used. The rate of strain may have some influence on the shear value. The results reviewed, however, have been performed at different sites and with different vane sizes. Consequently, no conclusions can be reached.

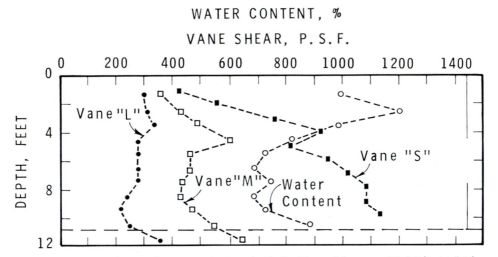

FIGURE 5.2   Vane shear and water content vs. depth. L, 4 in. × 8 in. vane; M, 2.8 in. × 5.6 in. vane; S, 2 in. × 4 in. vane

It appears from the review of vane shear testing at 16 sites that the vane shear test can be used to determine the shear strength of amorphous-granular and non-woody fibrous peats with sufficient accuracy for the results to be used for design purposes. It is obvious, however, that there is considerable need for further research concerning the possible relations between vane size, rate of strain, and shear strength. Until such research is carried out, it is tentatively proposed that, for design, vanes 4 in. (10.16 cm) in diameter and 8 in. (20.32 cm) long, employing a rate of strain of 0.1 degrees per second, be used. After a specific test procedure has been established, further research should be carried out to evaluate the influence of peat type on shear strength.

(2)   *Cone Penetrometer*
The particular cone penetrometer shown in Figure 5.4 was developed by the Waterways Experiment Station, U.S. Army Corps of Engineers, Vicksburg, Mississippi (U.S. Army 1959) to provide a measure of bearing strength and trafficability of mineral soils on which off-road vehicles travelled. Since 1962 this instrument has been applied to trafficability studies on organic terrain.

The tip of the cone penetrometer is slowly forced into the ground by pushing on the handle attached to the proving ring. The reading on the dial indicator, the

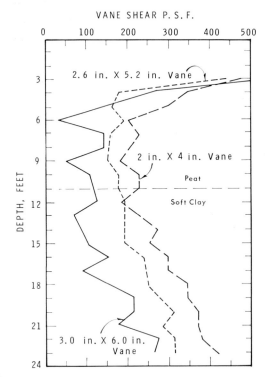

FIGURE 5.3   Vane shear vs. depth for three sizes of vane

cone index, is noted as the base of the cone penetrates the ground surface, and again at 3-in. (7.62-cm) penetration intervals up to 12 inches (30.5 cm) in depth and at 6-in. (15.25-cm) intervals for the remaining 24 inches (71.0 cm) of shaft.

The cone index, at present, can be correlated with vehicle performance on a "go–no-go" basis for 50 passes repeated over the same terrain (i.e., "go" condition means that the vehicle can complete 50 passes without becoming immobilized). For a given vehicle the cone index value corresponding to the line separating "go" from "no-go" conditions is constant and independent of the type of surface vegetation. Thus, once this value is known (as a result of tests) for a given vehicle, the cone penetrometer can be used to predict the muskeg areas that can be successfully crossed by the vehicle.

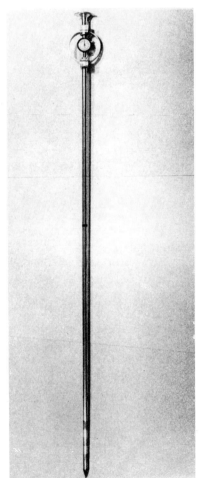

FIGURE 5.4   Cone penetrometer

While this instrument does not measure a well-defined physical property of peat, it is a highly portable and convenient tool in helping to assess muskeg trafficability in meaningful terms.

(3)   *Field Permeability* ∨

Measurements can be performed to determine the rate at which water will flow through peat under field conditions. Although work has been done (Boelter 1965) to determine horizontal and vertical permeability in the field, no significant difference has been detected between the two types of measurements. In general, well-humified peats have lower permeability than peats with a small amount of humification.

Common practice for measuring permeability in the field consists of sinking a pipe in the ground, removing peat and water from the pipe down to its lower end, and measuring the volume and time of the subsequent inflow of water. As an alternative, if the water table is fairly close to the ground surface, a simple well can be dug, a stake driven into the bottom of the well, and the water level marked on the stake. Some of the water is quickly pumped out of the well, and the distance from the original level to the new level, the volume of water removed, and the time required for the water still in the well to return to its original level are measured. These measurements can then be used to calculate the rate of flow of water through the peat.

Information obtained from permeability studies can be used in planning a proper drainage system design. This is a very important factor in road and railway construction on muskeg to avoid either flooding or excessive drying which could cause large settlements.

While the method of permeability measurement described above can provide good estimates, it would be presumptuous to say that the measurements actually correspond to a single linear flow rate of water through peat. When water is pumped out of the pipe or well, the water that flows in to replace it not only moves vertically up through the peat directly beneath the pipe, but flows in all directions through the peat surrounding the bottom end of the pipe. The rate of water flow at any point in the surrounding peat probably varies because of pressure differences occurring with changes in depth.

If a more precise determination of permeability in the field is required, it may be desirable to investigate the use of radioactive tracers dissolved in the water surrounding the well. If the flow velocities could be determined in this way, a more complete concept of peat permeability could possibly be formed.

(4)   *Density* √

Measurement of *in situ* peat density in the field can be accomplished with some inconvenience but with reasonable accuracy by using a nuclear (gamma) density meter.

In operation (Fig. 5.5) an access tube is inserted in the ground and a nuclear gamma source and a detector separated from the source by a shield are lowered into the access tube. Radiation is emitted into the peat surrounding the access tube and some of it, in an amount proportional to the peat–water mixture density, is reflected to the detector. The detector converts the reflected radiation into an electrical signal which is measured by a scaler on the surface.

An alternative method involves inserting two access tubes in the ground a known distance apart. The source is lowered in one tube and the detector in the other so that the attenuation of radiation transmitted through the peat can be measured. This attenuation is proportional to peat density. The advantage of this second method is that a narrower zone of peat is included in the measurement than is the case with the single pipe method.

In situ measurement of peat density is desirable since it is the only way in which the density of undisturbed peat can be ascertained. Any process of sampling (except frozen core samples) distorts or disturbs the peat and allows water to escape, so that measuring the density of the samples does not give the true values of the peat density in the deposit from which they were obtained.

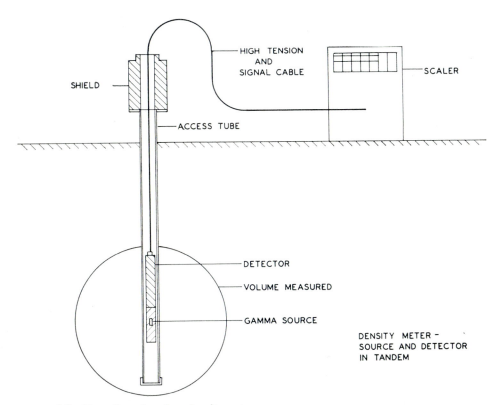

FIGURE 5.5   Operation of nuclear density meter

(5)   Water Content

The measurement of peat water content in the field can be achieved by using essentially the same techniques, and for the same reasons, as those described above for density measurement. The only difference is that for water measurement the gamma source is replaced by a source of fast neutrons.

In addition to being somewhat cumbersome to use, the neutron moisture meter has the disadvantage that trace elements such as iron present in the water (Burn 1966) can introduce intolerable errors in the water content measurements. There is some possibility that this disadvantage can be overcome with new instrumentation now being developed which responds to changes in the dielectric constant of soil–water mixtures resulting from changes in their water content.

Sometimes, when the peat is not extremely wet, a piston sampler can be used to obtain peat samples in which all the water is retained. The samples, or portions of them, can then be weighed, dried at an appropriate temperature, and weighed again to determine their water content (see Section 4.2(1)). Some error can occur through loss of water during the sampling operation, or during shipment to the laboratory. Also, during drying, loss of other volatile components of the peat can be misinterpreted as removal of water, thus causing a slight error.

(6)   Peat Depth

In order to determine the boundaries of a peat deposit, it is necessary to measure its depth. A common method is to manually force a steel rod into the ground. For a rod $\frac{3}{8}$ inch in diameter, forces of 100–400 pounds (45.4–181.4 kg) are necessary (Pihlainen 1963). This can give misleading results, however, if the peat is underlain by soft clay. Consequently, the most reliable results can be achieved by using a small-diameter (e.g. Davis-type) sampler or screw-type auger to obtain a depth profile. Power boring equipment consisting of an electric drill turning a sharpened coal auger has been used for frozen ground with some success.

Seismic methods have not been widely used for depth sounding in peat, probably because wave velocities are greatly dependent on water content and the presence of ice. Organic material attenuates seismic energy severely, thus further detracting from the usefulness of this method. Wave velocities in permafrost areas vary from 300 to 500 metres per second in "light muck" to 1300 to 3050 metres per second in "frozen muck" (Pihlainen 1963).

(7)   Test Fills

In road and railway construction over muskeg, it is sometimes desirable to construct test fills to evaluate deformation characteristics of the peat (Lea et al. 1964). In some instances the peat deposit is loaded to failure to determine the shear strength of the peat at failure. Load–settlement curves can be plotted which will aid in the prediction of settlement in the actual road or railway foundation; such curves can be used with more confidence than those based on laboratory consolidation tests alone. Instrumentation which is installed in the fill should include settlement plates to measure the settlement under load, piezometers to measure pore water pressure dissipation and inclinometers to measure lateral movement of the organic material under the fill. Reference is made to instrumented fills in Sections 6.5(4) and 6.16, and various types of instrumentation are illustrated in Figures 6.12, 6.13, 6.14, and 6.18.

5.3 SAMPLING

The type of samples obtained and the method used to obtain them depend on the type of peat structure, the water content, the use to which the samples will be put, and the type of sampler available.

In sampling peat, several difficulties arise from the fibrous structure of some peat types and the high free water content that often exists. When most samplers are inserted into the peat, the cutting edges of the sampler often do not sever the fibrous elements before they have been deflected away from their original positions. Thus, when the sample is extracted, it is so disturbed that it does not exhibit the same structure or physical properties as it did before it was removed from the ground. Furthermore, as samples of unfrozen wet peat are taken from the ground, a considerable amount of water often drains from the sampler and escapes. Therefore, if the samples are subsequently used in water content determinations, the results will be erroneous. In spite of these shortcomings, however, it is still useful to obtain samples to gain an appreciation of peat structure, humification, approximate water content and density, as well as to carry out strength and deformation tests.

Some samplers are of the core type which remove a cylinder of peat from the ground. Depending on the sampler used, varying degrees of disturbance are experienced by the peat. Other types of sampling involve excavation or cutting blocks from peat deposits. Brief descriptions of the most popular sampling methods follow.

(1) Hiller Borer
This sampler, illustrated in Figure 5.6, has been used for peat studies by palaeo-botanists, etc. for well over half a century. It consists of a chamber 50 cm (19.68 inches) in length and 2.8 cm (1.10 inches) in diameter with a longitudinal slot. The slot remains closed by a rotating sliding cover as the sampler is forced into the ground by rods attached to the upper end of the chamber. At the desired sampling depth, the sampler is rotated counterclockwise, thus opening the slot in the chamber, and a flange on the sliding cover scrapes the surrounding peat into the hollow chamber. The sampler handle is then rotated clockwise to close the chamber and the sampler is raised to the surface.

In very wet peat, water in the sample flows easily out of the sliding cover and tends to distort and mix the peat in the chamber. In drier peats, although the scraping action of the flange remoulds the peat horizontally, the relative vertical positions of the peat components are quite well preserved. Proper samples are difficult if not impossible to obtain where the peat contains non-humified woody components more than about ⅛ inch (31.8 mm) in diameter. Finally, if sand enters the clearance space between the sliding cover and the chamber, the sampler jams and must be dismantled and cleaned.

This is a useful tool for a quick field evaluation of peat structure and degree of humification since it is highly portable and can be carried and operated by one man. Because of the amount of disturbance to which it subjects samples, however, it is considered by some authorities to be unsuitable for obtaining samples for laboratory analysis.

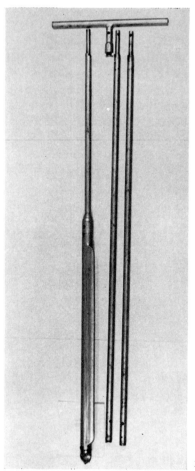

FIGURE 5.6   Hiller sampler

(2)   *Davis Sampler*

The Davis sampler (Fig. 5.7) has a retractable piston, really a loose fitting rod, inside a tube 15 inches (38.1 cm) long and ½ inch (1.27 cm) in diameter. The rod fills the tube while the sampler is inserted in the ground to the desired depth. The rod is quickly retracted until a spring catch at its lower end engages the top of the tube. The open tube is forced farther into the ground and then withdrawn, bringing with it a core of peat. To extract the sample, the spring catch is released and the core is extruded as the rod re-enters the tube.

Because of the large clearance between the tube and the rod, the Davis sampler does not successfully retain very wet peat samples, and because of its small diameter, it does not work well in peat containing woody fibrous elements. It does,

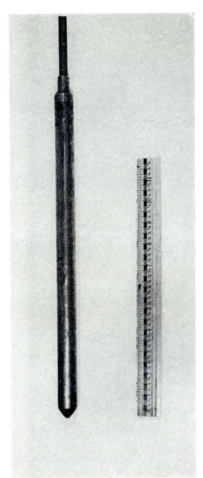

FIGURE 5.7   Davis sampler

however, have the advantages of light weight, easy operation, and simple maintenance, but the peat samples it procures are only of small value in carrying out preliminary field inspections of peat deposits. It should not, in general, be considered as adequate for peat sampling as a Hiller borer.

(3) *Piston-type Sampler*

A major improvement on the Davis sampler and, in some cases, the Hiller sampler, is the piston-type sampler shown in Figure 5.8. This particular version is a copy of a model that has undergone considerable development and use by the Geological Survey of Finland.

To operate the sampler, the piston is positioned at the lower end of the thin-walled stainless-steel-tubing cylinder and the shaft attached to the piston is used

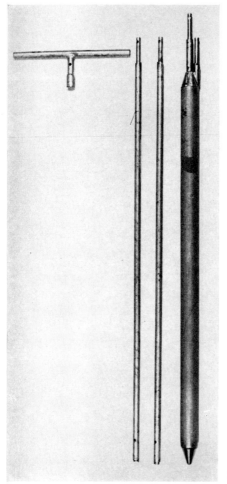

FIGURE 5.8   Piston sampler

to force the sampler into the ground to the desired depth. Then the piston is held stationary while the 2-inch (5.08-cm) diameter cylinder is forced farther into the peat by the separate rod attached to it. The sampler is removed from the ground by pulling on the piston shaft, and the peat core, about 2.5 feet (0.78 m) in length, is extruded by forcing the piston back down the cylinder.

Leather washers on the piston enable excellent suction to be set up within the cylinder and even very wet peat samples can be successfully obtained. In addition, very little distortion of the sample occurs and even most woody fibres are retained in the sample. The good results obtained with this sampler arise from the combination of shearing and suction actions.

For detailed examination of most peat deposits, this type of sampler is recommended for overall convenience, portability, and best retention of original peat structural integrity. Although in some ways similar in construction to the Shelby tube sampler, its piston design undoubtedly gives it superior performance in soft wet soil. Also it is usually constructed so that it can be used with the extension rods from a Hiller borer.

(4)   Steel Foil Sampler

This sampler has been used in geotechnical investigations of peat (Lea and Brawner 1959) and in some ways is similar to the piston sampler just described. A detailed description is given by Kjellman *et al.* (1950). Thin metal foils are individually coiled in storage spaces near the lower end of the cylinder and are attached to a loose fitting piston. As the piston remains stationary and the cylinder is forced into the ground, the foils unwind, surrounding the soil sample as it enters the cylinder. In this way sliding friction between the soil sample and the cylinder walls is avoided, thus preventing distortion of the sample. The sampler requires relatively heavy boring equipment for its operation, and because of the high precision required in its manufacture, it tends to be expensive.

(5)   Block Samples

Fibrous peats with medium to low water contents are amenable to being sampled in blocks from the walls of excavations in peat deposits (Fig. 5.9). Peat cutting tools such as those used in parts of Europe for digging peat for fuel might be used for this type of sampling. Alternatively, a combination of a flat spade, a large knife, and a fine-toothed saw have been used to obtain such samples.

(6)   Frozen-core Sampling

A somewhat cumbersome, but effective, method of obtaining undisturbed peat samples with fully retained water content is that of frozen-core sampling (Fig. 5.10). A pipe about 4 inches (10.16 cm) in diameter, with an obliquely cut lower end, is forced into the ground. A second pipe of larger diameter but with a constricted lower end is then forced into the ground so that it surrounds the first pipe and leaves a clearance space between the two pipes. Liquid air or a similar cold

FIGURE 5.9   Block sampling

liquid is poured into the clearance space and kept topped up for about half an hour. At the end of this period the peat in the inner pipe is completely frozen and remains in the pipe when the two pipes are extracted from the ground. The frozen sample can be removed from the inner pipe by pouring hot water on the outside of this pipe. This releases the core and it can be pushed out with a rod.

As might be inferred from the above description, this sampling method requires the support of off-road vehicles and equipment for forcing the pipes into the ground. When it can be used, however, frozen-core sampling provides the best possible peat samples for any type of detailed examination of peat structure.

(7) *Chain Saw*

When peat *in situ* exists in a frozen state, it is possible to obtain blocks of frozen peat with the assistance of an ordinary chain saw of the type used in forestry operations. No modification of the saw is necessary, and it is used in the normal way. There are obvious limitations to this method of sampling, however, and it is not likely to be as widely used as other methods.

(8) *Selection of Sampling Method*

It will be apparent from the above descriptions that the various sampling methods yield different results and have varying amounts of success depending on the

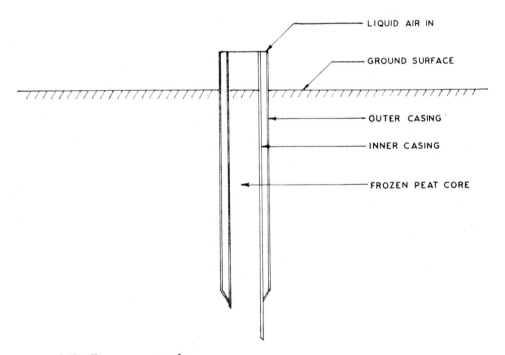

FIGURE 5.10   Frozen core sampler

sampling situation. It is suggested that the individual requiring samples decide what the samples will be used for and, bearing in mind the conditions under which they will be obtained, use the above set of descriptions to select the type of sampler which seems likely to achieve the most success.

(9)   *Care and Handling of Samples*

When samples are obtained from a test site, it is often necessary to preserve them as undisturbed as possible for future analysis. They should not be deformed or broken, all their water should be retained, and they should not be exposed to the air for long periods since this will promote decomposition of the peat. In spite of any precautions which may be taken, deterioration of peat samples can be caused by volume distortion from the presence of gas, loss of water during sampling or later handling, and disturbance of the sample as it is transferred from one container to another (Hillis and Brawner 1961).

These disturbances can be minimized, however, if certain precautions are taken. Samples from a piston borer can be carefully extruded into plastic bags or sleeves whose ends are then sealed with rubber bands or masking tape. These packages are then placed in cardboard mailing tubes for transporting and storage. If a supply of cylinders to fit the sampler is available, the cylinder full of peat can be detached from the sampler, and the ends securely sealed. The cylinder is then wrapped in thin plastic film of the household variety and finally in aluminum foil.

If air temperatures are below freezing or refrigeration facilities are available in the field, the cores can be left in the cylinder of the sampler for transporting and storing. To remove the frozen core, hot water is poured over the cylinder walls and the core is pushed out with a rod.

REFERENCES

ADAMS, J. I. 1963. The consolidation of peat – field and laboratory measurements. Ontario Hydro Res. Quar., Vol. 15, Fourth Quarter, pp. 1–7.

ANDERSON, K. O. and R. C. HAAS. 1962. Construction over muskeg on the Red Deer Bypass. Proc. Eighth Muskeg Res. Conf., NRC, ACSSM, Tech. Memo. 74, pp. 31–41.

ANDERSON, K. O. and R. A. HEMSTOCK. 1959. Relating the engineering properties of muskeg to some problems of fill construction. Proc. Fifth Muskeg Res. Conf., NRC, ACSSM, Tech. Memo. 61, pp. 16–25.

BOELTER, D. H. 1965. Hydraulic conductivity of peats. Soil Sci., Vol. 100, No. 4, pp. 227–231.

BRAWNER, C. O. 1957. Classification, laboratory testing and highway construction procedure for organic terrain. British Columbia Dept. Highways, Tech. Bull. No. 2, Victoria. 59 leaves.

———— 1959. Preconsolidation in highway construction over muskeg. Roads and Eng. Const., Vol. 97, No. 9, pp. 99–104.

BROCHU, P. A. and J. J. PARÉ. 1964. Construction de routes sur tourbières dans la Province de Québec. Proc. Ninth Muskeg Res. Conf., NRC, ACSSM Tech. Memo. 81, pp. 74–108.

BURN, K. N. 1966. The effect of iron in the determination of moisture content by the neutron method. Can. J. Earth Sci., Vol. 3, No. 1, pp. 129–132.

CADLING, L. and S. ODENSTAD. 1950. The vane borer: an apparatus for determining the shear strength of clay soils directly in the ground. Royal Swedish Geotech. Inst., Proc. No. 2.

HARDY, R. M. and S. THOMSON. 1956. Measurement of the shearing strength of muskeg. Proc. Eastern Muskeg Res. Meeting, NRC, ACSSM, Tech. Memo. 42, pp. 16–24.

HELENELUND, K. V. 1967. Vane tests and tension tests on fibrous peat. Proc. Conf. on Shear Strength Properties of Natural Soils and Rocks, Vol. 1, Oslo, pp. 199–203.

HILLIS, S. F. and C. O. BRAWNER. 1961. The compression of peat with reference to the construction of major highways in British Columbia. Proc. Seventh Muskeg Res. Conf., NRC, ACSSM, Tech. Memo. 71, pp. 204–227.

KJELLMAN, W., T. KALLSTENIUS, and O. WAGNER. 1950. Soil sampler with metal foils: device for taking undisturbed samples of very great length. Royal Swedish Geotech. Inst., Proc. No. 1.

LEA, N. D. and C. O. BRAWNER. 1959. Foundation and pavement design for highways on peat. Proc. Fortieth Convention Can. Good Roads Assoc., Ottawa, pp. 406–424.

———— 1963. Highway design and construction over peat deposits on the lower mainland of British Columbia. Highway Res. Board, Res. Rec. No. 7, Washington, DC, pp. 1–33.

LEA, N. D., C. O. BRAWNER, N. W. RADFORTH, and I. C. MACFARLANE. 1964. Prerequisites for design of engineering works on organic terrain – a symposium. Proc. Ninth Muskeg Res. Conf., NRC, ACSSM, Tech. Memo. 81, pp. 40–59.

MACFARLANE, I. C. 1960. Evaluation of road performance over muskeg in Northern Ontario. Proc. Sixth Muskeg Res. Conf., NRC, ACSSM Tech. Memo. 67, pp. 38–48.

———— 1964. Muskeg field program, Mer Bleue peat bog. Personal correspondence.

MICKLEBOROUGH, B. W. 1961. Embankment construction in muskeg at Prince Albert. Proc. Seventh Muskeg Res. Conf., NRC, ACSSM. Tech. Memo. 71, pp. 164–184.

PIHLAINEN, J. 1963. A review of muskeg and its associated engineering problems. U.S. Army Materiel Command, Cold Regions Res. and Eng. Lab., Tech. Rept. 97, Hanover, N.H., 56 pp.

RADFORTH, N. W. 1961. Distribution of organic terrain in Northern Canada. Proc. Seventh Muskeg Res. Conf., NRC, ACSSM Tech. Memo. 71, pp. 8–11.

RIPLEY, C. F. and C. E. LEONOFF. 1961. Embankment settlement behaviour on deep peat. Proc. Seventh Muskeg Res. Conf., NRC, ACSSM Tech. Memo. 71, pp. 185–204.

TESSIER, G. 1966. Deux exemples – types de construction de routes sur muskegs au Québec. Proc. Eleventh Muskeg Res. Conf., NRC, ACGR Tech. Memo. 87, pp. 92–141.

———— 1967. Etude complémentaire sur les surcharges dans les savanes de Napierville. Proc. Twelfth Muskeg Res. Conf., NRC, ACGR Tech. Memo. 90, pp. 107–129.

6 Road and Railway Construction

Chapter Co-ordinators *
C. O. BRAWNER
G. TESSIER

6.1 INTRODUCTION

For many centuries the conventional approach where organic terrain occurs has been to construct roads on stable ground around the muskeg. Where the peripheral distance is great, the incentive to cross the muskeg is increased.

The earliest known roads built in muskeg have been found in the Brue Valley near Somerset, England. Dewar (1962) reports excavating peat and finding old roadways constructed of planks. These were tied together longitudinally at each end with stakes driven vertically every 10 feet (3.05 m) along each side to hold them in place. Figure 6.1 illustrates one of these typical roads which carbon dating has indicated to be between 4000 and 4800 years old.

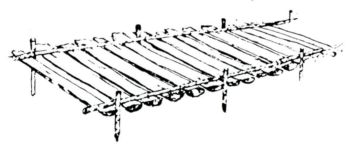

FIGURE 6.1   Plank road in muskeg in Brue Valley, Somerset, England. Carbon dating has indicated that this road is between 4000 and 4800 years old

Very few changes took place until the current century. Surface corduroy or planking provided too rough a surface for early automobile traffic. As a result, soil was spread on top of the corduroy to improve the riding qualities. The corduroy provided a certain amount of buoyancy, spread the weight of the road more evenly

*Chapter Consultants: C. T. Enright – Secondary and Access Roads; F. L. Peckover – Railway Construction.

over the muskeg, and helped to prevent fill material from penetrating the surface of the muskeg. Wire mesh, dried bundles of peat, and bundles of straw have also been used as a base for fill.

As the frequency and weight of traffic became greater, the road standards increased. Since the composition of peat and the depth of the deposits were variable, differential settlement under this fill was frequently a serious problem.

In the nineteen thirties, the Michigan State Highway Department pioneered the method of excavating and displacing the peat (Cushing and Stokstad 1934). In the shallower deposits, excavation was frequently performed with draglines or shovels and the excavated area backfilled with inorganic soil, preferably sand and gravel. In very deep deposits, partial excavation and backfill were frequently used. This technique did not prove too successful because of major settlement and it led to the development of displacement methods of construction. With gravity displacement, the leading edge of the fill was built up to such a point that overloading occurred. The fill material settled, displacing the underlying peat. To assist this process, water jetting was occasionally used to reduce the shear strength of the peat and facilitate displacement. Blasting was also used to displace the peat. Two methods, the underfill and toe shooting techniques, were developed (Parsons 1939).

In the mid fifties, efforts were directed towards construction without removal of the peat. Brawner (1958) reports the results of an experimental road section to test preconsolidation of peat. With this method, a load in excess of that finally proposed is placed to induce settlement. When the rate of settlement decreases significantly, the excess weight is removed. Subsequent settlement is usually small. This procedure has been used successfully on more than ten major projects in Canada. In recent years, several modifications of this method have been developed. Lea and Brawner (1963) report the use of sawdust as lightweight fill to reduce the ultimate load on the peat and to act as a weightless spacer where excessive settlements were expected in the peat. Flaate and Rygg (1964) in Norway outlined a similar approach but used timber as a base for the sawdust.

Where muskeg is deep for short distances or where heavy loads are required, bridges on piles have been used. A summary of the methods of road construction over organic terrain that have been used to date is shown in Table 6.1 (cf. Mac-Farlane 1956).

6.2   PRELIMINARY DESIGN CONSIDERATIONS

The method of construction selected should be the one that provides the desired standard of road to serve the user efficiently at the lowest possible cost. Numerous methods are available, each of which has explicit applications and limitations.

The first factor that should be evaluated is the standard of highway to be constructed. If only local traffic is to be served, considerable settlement, undulation, distortion, etc. can be tolerated. Under these conditions, construction of the road

TABLE 6.1
Road construction methods over organic terrain

Floating the road	Excavation and replacement	Displacement	Preconsolidation	Special methods	Relocation
Spread foundation (thin wide fill)	Total excavation and replacement with stable fill	Gravity surcharge method with or without water jetting	Standard fill	Bridging on piles	
Corduroy construction (fill over timber or brush)	Partial excavation and replacement with stable fill	Blasting Underfill method Toe shooting method	Lightweight fill	Vertical sand drains	
Lightweight materials (fill over peat, straw, sawdust)				Natural drainage	
Reinforced mattress (fill over wire mesh war surplus landing mats)				Chemical stabilization	
Lateral support (berms, rock-filled side ditches)					
Frozen muskeg (place fill on frozen muskeg)					

on top of the muskeg will usually be adequate (see Part II of this chapter). With a gradual increase in the amount and weight of traffic, the stability can be increased by additional surfacing material. On the other hand, where the road handles a considerable volume of traffic, such as a primary route or freeway, a design must be selected which will provide a high standard of grade and alignment. A reasonably maintenance-free design life is an important consideration. For example, a low standard road might be designed for 5–10 years, a primary route for 15 years, and a freeway for 30 years.

For roads of low standard, the availability and cost of construction materials such as corduroy, sawdust, gravel, etc. will usually dictate the method of construction. In very soft muskeg, it may be necessary to place corduroy or sawdust before gravel can be placed. The thickness of the riding surface will depend on the loads to be imposed. Twelve inches (30.5 cm) of gravel may be adequate for automobiles, whereas 3–4 feet (0.91–1.22 m) will usually be required for heavy trucks.

For primary roads or freeways, many factors will influence the method of construction. These include the characteristics of the peat and underlying soil, construction materials, equipment required, available time, location of structures, and drainage requirements.

(1) Muskeg and Subsurface Soil Conditions
If the peat is shallow, it is desirable to excavate it completely and replace it with sand, gravel, or other stable materials. The minimum depth to which excavation is economical will depend on the cost of excavation, disposal, and backfill material. For deeper deposits, excavation becomes very expensive and alternative procedures such as displacement or preconsolidation should be considered.

Where peat is underlain by soft clay, a quite common condition, stability and the rate of settlement may be major problems. Where extra right-of-way is available, counterbalancing berms may be employed. Alternatively, lightweight fill construction may be utilized. If the settlement in the clay will occur too slowly, vertical sand drains, preferably installed by a non-displacement method, may provide the desired increased rate of consolidation. It should be noted that vertical sand drains have frequently been unsuccessful in peat (see Section 6.5(5)). Prior to the detailed consideration of such specialized techniques, the subsurface materials should be investigated thoroughly. Where primary highways will cross large muskeg deposits and displacement or preconsolidation are being considered, full-scale field tests are recommended prior to construction.

Where soft clay underlies the peat, complete excavation and replacement of the organic material may create a more critical stability and settlement problem than does preloading, since the load on the clay will be increased. This emphasizes the need to investigate the materials that underlie the peat.

(2) Construction Materials and Equipment
In some areas stable replacement materials are scarce or must be hauled long

distances. Preconsolidation requires less fill than excavation or displacement and should be considered.

Occasionally, the availability of construction equipment will have a bearing on the method of construction selected. Where the peat to be excavated is moderately deep and is very soft, it may have to be handled several times in order to move it far enough from the cut so it will not flow back into the excavation. On the Deas Freeway south of Vancouver, BC, draglines were required to sidecast the material 150 feet (45.7 m) from the cut. This is obviously expensive. Figure 6.2 shows typical dragline excavation near Kelowna, BC.

Where the project is near a river or large body of water, the possibility of dredging and pumping the fill material should be considered. Granular fill can be pumped economically for distances exceeding 2 miles (3.22 km).

FIGURE 6.2   Excavation of peat by dragline and replacement with sand fill near Kelowna, BC

FIGURE 6.3   Differential settlement of highway grade constructed by floating the road on the muskeg near Ladner, BC

(3)   Time

Climatic, physical, economic, or political factors may impose construction deadlines. Typical situations may include the necessity to curtail operations on a project during the rainy season or cold weather. Construction employing excavation or displacement methods is usually faster than preconsolidation and, if time is a critical factor, it may be necessary to adopt the former method even though the cost is greater.

If the peat is very deep and the grade is required at an elevation relatively high above the original ground, preloading may not be practical because of the time required for the necessary settlement to take place. In addition, stability may require stage construction.

(4)   Adjacent Structures

It may be necessary for highways to be located in muskeg adjacent to buildings, railways, or other highways. Excavation may result in lateral movement into the excavated area with subsequent damage. Near Maillardville, 30 miles (48.3 km) east of Vancouver, BC, the Burnaby Freeway parallels the Canadian Pacific Railway. Excavation in this area would have resulted in substantial movement of the railway. If surcharging is employed and the depth of soft compressible peat and mineral soil is large, settlement may occur for some distance from the toe of the fill and thus affect adjacent structures.

Occasionally bridges are required in muskeg areas. Approach grades 10–30 feet (3.05–9.15 m) high may be necessary. Detailed consideration must be given to selecting the best method of constructing the bridge approaches. Excavation of the peat and backfilling with stable material may improve stability and reduce the time required for construction if the stratum underlying the peat is stable. If it is very soft, stability may be adversely affected. Where time is not a prime consideration, preconsolidation may be possible. Detailed testing to evaluate settlement and stability are required, however, before a decision can be reached. If the abutments are to be founded on piles, negative skin friction must be considered in the design. Differential settlement can be expected between the bridge abutment and approach fill. If the abutment is founded on spread footings in a granular fill approach, the effect of settlement on the structure must be evaluated. In this case, it may be necessary to construct the abutments high enough that they will settle to grade later. Furthermore, settlement of the abutment will usually be associated with a backward tilting of the abutment. Bridge seat and expansion joints must be designed to accommodate this.

(5)   Drainage Requirements

Drainage design and control is straightforward if the peat is to be excavated and replaced. If preconsolidation is employed and considerable settlement is expected, culverts must usually be placed prior to construction, dug up following removal of surcharge, and reinstalled to proper grade. If the expected settlement is less than 2–3 feet (0.61–0.91 m), however, the culverts can normally be installed high enough to allow for the future settlement.

(6) Construction Problems and Control

Construction control on highways built in muskeg is generally more difficult than on those constructed in inorganic soils or rock. This is possibly due in large measure to the limited experience of engineers and contractors in dealing with muskeg rather than to technical difficulties. Each of the methods of construction requires certain controls, some simple, others quite complex. The most significant factors to be emphasized and controlled are discussed in detail in Section 6.5.

6.3 EVALUATION OF ECONOMICS

The fundamental requirement of the highway engineer is to provide a structure with the optimum combination of economy, practicability, and serviceability compatible with the tax dollar generated by the route under consideration.

Where muskeg is prevalent, the most important question which usually develops is the choice of location – to go around or through the muskeg? To the layman, highway costs are often associated only with construction costs. In reality, however, there are at least six other factors which bear major economic comparison. These include right-of-way costs, design and engineering costs, safety and accident potential, local and regional development benefits, road user benefits, and future maintenance.

(1) Right-of-Way

In populated or industrial areas the cost of right-of-way for roads around the muskeg may be very high, whereas the price of property in the muskeg often will be low. On the Burnaby Freeway through Vancouver, BC, for instance, right-of-way costs immediately adjoining the muskeg were as high as $30,000–$40,000 per acre, whereas in the muskeg areas the price ranged from $100 to $2,000 per acre. A further factor to be considered is the comparative loss in property tax to cities, municipalities, governments, etc. resulting from the purchase of right-of-way. In the majority of instances, right-of-way costs and tax losses will be much less in muskeg areas.

(2) Design and Engineering Costs

Construction problems are frequently more critical in muskeg areas. More care, detail and, subsequently, expense is to be expected in the design, surveys, investigations, and construction control. In addition, the length of the construction period in muskeg may be longer than normal, particularly if preloading is used or if very soft clay, which contributes to a stability problem, exists. Often, special instrumentation may be required and this cost must be included.

(3)   Safety and Accident Potential

Normally, specific design limitations for maximum curvature and grade, minimum sight visibility, pavement width, and shoulder width exist for the highway to be constructed. By crossing directly over the muskeg, straighter and safer alignment and greater sight visibility usually can be obtained. In addition, muskeg provides a natural tendency for limited access, with few local roads entering the highway.

(4)   Local and Regional Development Benefits

The construction of highways plays a major part in residential, tourist, recreation, commercial, agricultural, and industrial development of any area. Each comparative highway location must be considered in the light of the relative benefits of both present and potential development.

(5)   Road User Benefits

It is often desirable to construct a highway between two points by the shortest possible route. The savings to the road user in time and motor vehicle operating costs can be substantial. For example, assuming that a road through muskeg would be shorter by 0.4 miles (0.64 km) than a road around the muskeg, if 4000 vehicles per day use the route and the average operating cost per vehicle per mile is 8¢, a yearly benefit of $0.4 \times 4000 \times 365 \times 0.08$ or \$47,720 is indicated. This does not include the benefit gained by the saving in time. Taken over a 10–20 year period, the road user benefits can be substantial.

   In some areas, muskeg is adjacent to hills. Positive grades to bypass the organic terrain may be required and the effect of these on the operating and time costs of commercial vehicles is important.

   It is possible that shorter routes may not provide local access previously available. In this case, the cost of providing new access must be included in the economic study.

(6)   Construction

The estimation and comparison of construction costs will depend on many factors. For routes around the muskeg, the depth and width of cut, height of fill, rock and soil type, soil characteristics and quantity, haul distances, and drainage requirements must be considered. For roads over muskeg, the structural requirements of the roadway, depth and strength of the peat and underlying soil, settlement characteristics of the peat and underlying soil, presence of adjacent structures, drainage requirements, availability of stable granular fill or replacement material, construction timing and control, construction equipment required, etc. influence the final decision.

Evaluation of the location, design, and costs should always be preceded by an adequate soils investigation.

(7)   Maintenance
An assessment of future maintenance problems and costs is desirable if a realistic economic picture is to be obtained. Roads around muskeg often will create more costly snow removal, drainage and ditch maintenance, slope stability, etc. Furthermore, retaining devices such as guard-rails may be required. In muskeg the most frequent maintenance factors required are relevelling a non-uniform grade and maintaining drainage through culverts.

In order to make a rational evaluation and comparison of the costs of constructing roads through muskeg, it is necessary to investigate many factors other than construction costs. It is quite common in fact, for construction costs through muskeg to be greater than those for bypassing muskeg. A comprehensive economic evaluation of all possible routes, however, may indicate that the route through the muskeg is the least expensive.

6.4   PRELIMINARY SITE APPRAISAL

A preliminary site survey must be undertaken before feasibility studies and route selection studies can be finalized. Two of the most useful techniques for site surveys are airphoto interpretation and aerial photogrammetric mapping (Brawner 1960a). Aerial surveys provide information without field inspection and provide a comprehensive view of the terrain and regional development.

The technique of airphoto interpretation is well established. Stereoscopic study of airphotos reveals the shape of the land surface and the position of man-made features. By assembling a mosaic of several flight strips, it is possible to study all possible highway routes. By evaluating such features as topography, drainage, colour tone, vegetation, land use, stream profiles, and special markings it is possible for the interpreter to delineate land forms, soil types, peat deposits, aggregate sources, etc.

Photogrammetry is the science of obtaining horizontal and vertical measurements from aerial photographs. Topographic maps with a 2-, 5-, or 10-foot (0.61-, 1.52-, or 3.05-m) contour interval covering a wide band of terrain can be obtained much more economically in lightly vegetated areas by this method than by conventional field surveys.

Airphoto interpretation and photogrammetric mapping provide much basic information for preliminary highway location surveys in areas where organic terrain occurs. An experienced interpreter can estimate with reasonable accuracy the type, characteristics, and depth of the peat and predict the engineering problems to be expected (see Chapter 3). Airphoto surveys are recommended for all preliminary route location studies.

If the preliminary survey indicates the feasibility of a route or routes through muskeg, a field investigation to determine accurately the physical properties and depth of the peat is desirable. This involves determination of the depth of the muskeg and the contour of the firm bottom to indicate the best location. The information can be obtained by forcing a steel sounding rod by hand through the peat. Notes are made to include the pressure required to force the rod into the ground at various depths. By drilling the occasional test hole at sounding points, the sounding rod pressure can be correlated with the general soil type and the correlation extended for all sounding points. This procedure is sufficiently accurate to provide the data required to select the final alignment.

On the final location, soundings are recommended every 25–50 feet (7.62–15.25 m) along the centre line, with offset soundings every 50–100 feet (15.25–30.5 m). Test holes should be drilled every 100–300 feet (30.5–91.5 m) along the centre line to obtain samples for the identification and determination of physical and mechanical properties. If settlement computations are required, undisturbed samples are required for consolidation tests.

To estimate the height of fill that can be placed initially, field vane shear tests at every 1 foot (0.305 m) increment of depth are recommended. These are usually spaced 300–500 feet (91.5–152.5 m) apart along the centre line.

6.5   DESIGN AND CONSTRUCTION

(1)   *Construction Directly on the Muskeg*

If first cost is the main consideration, the most inexpensive method of constructing roads is to place fill directly on top of the muskeg. This is frequently referred to as "floating the road." The fill, preferably sand and gravel, may be placed alone or in conjunction with corduroy, brush, sawdust, wire mesh, dried bundles of peat, bundles of straw, etc. With this method of construction, serious differential settlement frequently occurs (Fig. 6.3). If good riding characteristics are required, extensive maintenance will probably be necessary.

Construction directly on muskeg is most applicable where light traffic or low speed limits are expected, where the muskeg is very deep, or where the proposed highway is adjacent to a structure that would be disturbed by other methods of construction.

Prior to design, a field survey should be made to determine the continuity, thickness, type, and relative strength of the surface mat. The depth of the peat, the peat type, and engineering properties such as water content, compressibility, and vane shear strength should be determined and the mineral soils underlying the peat should be sampled. If soft or sensitive soil is encountered, the shear strength must be determined in order that the stability under the fill load may be assessed.

The minimum thickness of fill and surface course required is that which will carry traffic for 10–20 years without pavement failure. Where the surface mat is of moderate strength, at least 18 inches (45.7 cm) of pavement (granular fill plus asphalt) should be used. For heavy traffic, 4 feet (1.22 m) of pavement structure is the minimum recommended (Brawner 1960b) as indicated in Figure 6.4.

Stability analysis using vane shear tests in the peat and unconfined compression or triaxial tests on any underlying soft soils will indicate the general order of fill that can be applied safely. Consolidation tests on the peat will indicate the order of settlement to be expected.

Where the peat and underlying soil have a shear strength in excess of 120 pounds per square foot (586 kg per sq m), 4 feet (1.22 m) of fill can be placed safely. On the Burnaby Freeway in Vancouver, BC, shear strengths of 80–100 pounds per square foot (391–488 kg per sq m) were encountered and a failure occurred when 3 feet (0.91 m) of fill had been placed (Lea and Brawner 1963).

The amount of settlement in peat may exceed the thickness of the fill applied. On the project noted above, it was computed that 3 feet (0.91 m) of fill would cause a settlement of 9 feet (2.75 m). Under such circumstances it is necessary to place the fill in conjunction with a lightweight material such as corduroy, wood shavings, sawdust, etc. Sawdust has been found to be most successful in British Columbia, where it is relatively inexpensive. It is very easy to handle, compacts readily, and provides an excellent roadbed for construction equipment under all climatic conditions. It is recommended only where it will settle below the water table, to ensure that deterioration or spontaneous combustion will not occur.

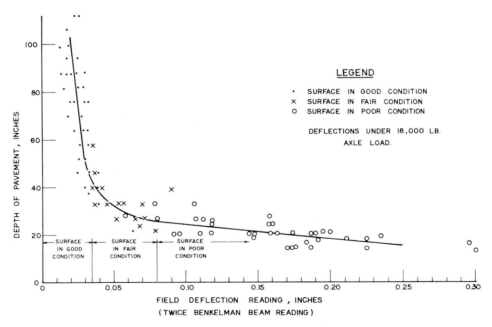

FIGURE 6.4   Depth of pavement on peat vs. field deflection reading

Prior to construction, all woody vegetation should be cut off at ground level, with the smaller cuttings placed as a mat on the surface. Under no circumstances should the surface mat of muskeg be disturbed. Trees or stumps should not be burned on the muskeg because the fires may spread and continue for many months or years. Figure 6.5 shows a typical cleared portion of the right-of-way through muskeg.

If the muskeg is very weak and has little or no surface mat, it may be necessary to place corduroy or use fascine construction. Logs placed side by side or criss-cross are most common. Other materials used are planks, brush, wire mesh, dried peat, bundles of straw, sawdust, or wood chips. Their purpose is to spread the load of the fill, provide some buoyancy, and prevent the fill from sinking into the muskeg.

Granular soils are preferred for fill on top of muskeg. They are easy to place, grade, and compact and they provide better subgrade support. Rock has a tendency

FIGURE 6.5   Clearing of the right-of-way completed prior to the use of preconsolidation construction on the Trans-Canada Highway near Chilliwack, BC

FIGURE 6.6   Gravity displacement of organic lake-bed soil near Revelstoke, BC

to sink into the muskeg, and clay inhibits drainage. At least 3 feet (0.91 m) of granular fill on peat is required to support loaded construction vehicles. Where the fill thickness on the final section is proposed to be less than this, a central fill at least 3 feet (0.91 m) thick can be placed by end dumping and then spread to final thickness by a small caterpillar tractor. Compaction is best achieved by using small vibratory rollers.

If the fill load is expected to cause a shear failure, a counterbalancing berm can be placed. The berm should be about half the thickness of the main fill and about 20 feet (6.1 m) wide. Standard soil mechanics procedures can be used to determine the berm dimensions. An alternative method provides lateral support by excavating a trench along the fill and filling it with sand, gravel, or rock.

Cronkhite (1967) has described a procedure that has been successful in winter construction for secondary standard highways. When the peat has frozen to develop sufficient strength to carry heavy construction equipment, fill is placed directly on the peat, spread, and compacted rapidly before the fill has a chance to freeze. This method has overcome several normal construction problems.

(2)   *Excavation and Replacement by Stable Fill Material*
Until about 1957 the most obvious approach to the problem of construction over muskeg was to excavate the peat and backfill with stable fill, preferably sand and gravel. This method is still considered by many engineers to be the most dependable way of constructing permanent roads over organic terrain. Provided that all the peat can be excavated, a stable roadbed exhibiting negligible settlement can be constructed. Another advantage of total excavation is that the fill is thick enough to support an asphalt or concrete surface.

In general, however, total excavation is only economically feasible for shallow depths of peat. Opinions differ regarding the maximum economic depth, which depends upon the type of peat, excavation equipment available, length of the deposit, and cost of the fill material. The economical depth generally varies between 6 and 12 feet (1.83 and 3.66 m).

Several problems frequently develop. It is not uncommon for local pockets of peat to remain unexcavated. The excavation may partially fill with water and hamper the view of the operator of the excavation equipment. These local pockets subsequently result in differential settlement of local areas of the roadway.

If the peat has a very low shear strength, which is more common in amorphous-type peat, the side slopes of the excavation may fail. This can result in a significant increase in excavation quantities.

The excavated peat is usually sidecast beyond the edge of the excavation. If the peat is soft it is not unusual for the weight of the sidecast material to create a slide back into the excavation. Where peat was excavated for construction of the Deas Freeway, south of Vancouver, BC, it was necessary to sidecast the material up to three times so that the weight of the excavated peat would not cause lateral displacement back into the excavation. Where a highway is proposed adjacent to existing structures the possibility of the excavation undermining those structures must be evaluated.

Total excavation of peat and replacement with stable fill are applicable chiefly where peat deposits are shallow, where they are not located adjacent to a structure, and where there is an abundant supply of low-cost granular fill available. Where traffic and vehicle considerations require rigid concrete pavements, it is highly desirable that differential settlement be kept to an absolute minimum and complete excavation of the peat is normally recommended.

The success of highways constructed in muskeg by excavation and replacement will depend to a great extent on the detail of the soil investigation. Once a decision has been made for complete excavation, it is desirable that a combined boring and sounding programme be carried out over the entire length of the project to be excavated. Shear strength data obtained with field vane shear equipment, particularly in amorphous-granular or fine-fibrous peats, can be useful in estimating the steepness of the slopes to which the peat will stand. For depths up to about 5 feet (1.53 m) slopes of 1 horizontal to 1 vertical are frequently used. Beyond a depth of 8 feet (2.44 m), the side slopes vary considerably and may be as flat as 4 horizontal to 1 vertical.

If it is possible to establish natural drainage in side ditches at a depth several feet below the existing surface of the muskeg, it is not necessary to backfill the entire depth of the excavation. On the Deas Freeway the final grade was established 6 feet (1.83 m) below the average top elevation of the muskeg. This reduction in grade saved about 500,000 cubic yards (382,000 cu m) of granular backfill. As the vegetation began to grow, the slopes soon took on a pleasing appearance.

Two basic methods of construction are used. In the first method draglines operate off the end of the fill, excavating immediately in front of it. This provides a stable footing for the dragline. The fill is placed by end-dumping from trucks and pushed into the excavation with caterpillar tractors. This method has the advantage that the excavation is left open for a minimum period of time and the side slopes will normally stand at a steep angle. The draglines can frequently cast backwards at 45 degrees so that the weight of the excavated material will not aggravate the stability of the excavated slope.

The second method is used where granular material can be obtained economically by pumping from a nearby source. When the excavation is completed, granular material is pumped into the trench in one continuous operation. Where the proposed construction is adjacent to navigable water, it has been found feasible to excavate the peat by dredging and then to backfill the trench by a normal trucking operation or by pumping (Engineering News-Record 1945).

If it is possible for the slopes of the peat to be excavated fairly steeply, the lateral load applied by the weight of the granular fill will cause some horizontal compression and movement of the road shoulder. As a result a shoulder width of at least 6 feet (1.83 m) and preferably 10 feet (3.05 m) is normally recommended with this type of construction.

A modification to complete excavation which has been used in peat over about 10 feet (3.05 m) in depth is to excavate a portion of the peat and backfill with stable material. This is often described as "partial excavation." The general procedures for design and construction combine those used for floating the road and for complete excavation. The amount of settlement that will occur, however, can be large. The weight of the fill placed is usually considerably in excess of the weight of the peat excavated and this increase in weight contributes to the increase in settlement. The weight can be reduced by using some lightweight fill. If the physical characteristics of the peat differ and the depth of the deposit changes over relatively short distances, moderate to severe differential settlement which will continue for several years can be expected. With this method, it is desirable that the construction of the pavement surface be delayed as long as possible to allow most of the settlement to take place.

(3)   *Displacement Methods*

A   Displacement by Gravity

Gravity displacement of peat has been carried out on many projects, particularly in the United States. As the fill is constructed across the muskeg, its frontal face is advanced with a "V" shape and surcharged sufficiently so that a shear failure takes place, displacing the peat laterally. A typical example near Revelstoke, BC, is shown in Figure 6.6. This technique is most successful where the peat is 10–20 feet (3.05–6.1 m) in depth and has a low shear strength. If the peat is relatively stiff and difficult to displace, additional weight may be obtained by continuous watering of the fill.

A more frequent technique for reducing the shear strength is jetting. Early attempts at jetting required that the fill be placed first and that jets be forced down through the fill into the peat below. The additional weight of the fill and consolidation of the peat made it difficult to bring the peat to a fluid consistency. Greater success is most often achieved with water pumped into the peat before the fill is placed, during placement, and as long thereafter as the fill continues to settle. Occasional borings may be necessary to indicate when extra jetting is necessary. The spacing of the jets depends on the type and depth of peat. Experience in Michigan has indicated that for best results jets should be spaced 10–25 feet (3.05–7.62 m) apart over the entire area of the proposed fill (Cushing and Stokstad 1935). Clean sand and gravel fill is required. Clayey soils will tend to become fluid and to displace laterally. Centrifugal pumps with a capacity of 500 gallons (2272 litres) of water per minute operating at a 250-foot (76.3-m) head and 1450 r.p.m. can operate 12 to 15 jets per minute.

Care is required with jetting and gravity displacement techniques since pockets of peat, which will lead to future settlement may be left. As a result of the increased water content of the peat which is displaced laterally, the weight of the fill may cause considerable lateral compression, with the result that the shoulders may settle and move horizontally at a gradually decreasing rate for many years.

B   Displacement by Blasting

Displacement of peat beneath embankments has been successfully accomplished by blasting. Two techniques are in common use, underfill blasting and toe shooting.

In the underfill technique, the surface mat is broken by mechanical means or by light dynamite charges. Fill is then deposited on the surface of the muskeg and explosives are placed beneath the fill at or near the bottom of the peat deposit. The explosives may be placed in holes created by water jetting or by the use of casings driven through the fill and muskeg.

The location and amount of explosives used will depend upon the depth of the peat, its shear strength, and the depth and width of the fill. The usual procedure is to place one or two rows of charges along the centre line of the embankment and a further row of charges under the toe of each side of the fill. The outside charges are detonated about 25 milliseconds prior to the central charges. The explosion displaces the organic material under the fill and creates a cavity beneath the embankment, causing it to settle rapidly. Figure 6.7 shows the typical method of underfill blasting.

Forty to sixty per cent gelatine dynamite has been used most frequently. Spaced in three or four rows under the fill, one pound of explosive will usually displace 2–3 cubic yards (1.53–2.29 cu m) of peat (Jeffries 1936; Parsons 1939). The size of the charges ranges between 20 and 40 pounds (9.07 and 18.14 kg) of explosive. Displacement using the underfill method has been successful in peat up to 30 feet (9.15 m) deep.

In the toe shooting method, the peat is displaced by blasting ahead of the advancing fill. Additional fill is then pushed into the cavity left by the blast and the fill is advanced with a "V" point displacing the peat and developing a wave in front of the fill. The front face of the fill is surcharged and dynamite charges are placed around the toe of the fill near the bottom of the peat. These charges are detonated and the peat is again blown out ahead of the fill (Figure 6.8). This method has been used successfully for depths of soft peat up to 20 feet (6.1 m) deep.

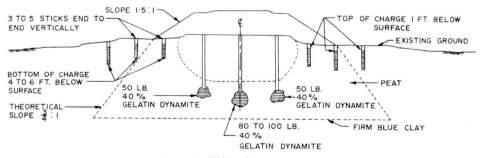

FIGURE 6.7   Peat displacement by underfill blasting

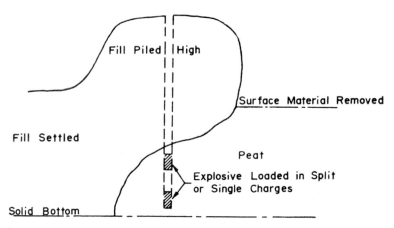

FIGURE 6.8   Peat displacement by toe shooting

In deeper deposits, instead of the entire charge being placed at the bottom of the organic deposit, separate sticks of explosive are tied at various points on a pole whose length is approximately equal to the depth of the peat. Several of these torpedo-type charges are placed vertically in the peat under the fill and detonated. The best success is obtained by using a 20–25 millisecond delay with the top charges detonated first.

Free-draining sand or gravel should be used wherever possible for fill. If these materials are not economically available, clays, silts, or till may be used, but large differential settlements, some of which may take several years to occur, can be expected.

The major problem experienced with displacement by blasting is that all of the peat may not be displaced in which case some differential settlement can be expected. Settlement and horizontal movement of the shoulders also occur frequently.

(4)   Preconsolidation

Preconsolidation to increase the shear strength and to reduce the long-term settlement of soft soils has been used on many projects where soft mineral soils were encountered. In 1958, the technique was first applied to construction in muskeg in Canada (Brawner 1958, 1959). The technique of preconsolidation requires that a load is placed in excess of that which will be finally carried by the soft stratum, and allowed to settle until the ultimate settlement that would occur under the final load has been reached. The excess load or surcharge is then removed and the construction is completed.

The principle of preconsolidation is illustrated in Figure 6.9 (see also Section 4.4(5) and Fig. 4.10). Curve 1 shows the time–settlement relation for adding

6 feet (1.83 m) of fill to a peat deposit 7 feet (2.13 m) in depth. Settlement after 1 year is 2.8 feet (0.855 m) and after 25 years is 3.5 feet (1.07 m). The 25-year settlement of 3.5 feet (1.07 m) can be produced during the construction period by placing an 11-foot (3.36-m) fill for approximately one month (curve 3). When the excess 5 feet (1.52 m) of fill is removed, future settlement is theoretically negligible.

The main advantages of the preconsolidation method of construction are as follows: reduced fill material is employed, no peat excavation which may undermine adjacent structures is necessary, and disposal areas for the peat are not required.

Some potential problems exist with the preconsolidation method. If more than 3–8 feet (0.91–2.44 m) of fill is required to gain full surcharge height, the effect of the fill on stability must be ascertained. This problem is potentially most critical where soft clay underlies the peat. If moderately high fills are required, stage construction may be necessary to build the fill to full surcharge height. In this case, the length of time required to complete construction could be excessive.

Considerable settlement can be expected. The amount of this settlement must be known with reasonable accuracy during the design stage. Occasionally, the peat may be so soft that the amount of settlement will be somewhat more than the thickness of the fill. On the Burnaby Freeway in Vancouver, for example, it was estimated that 9 feet (2.75 m) of settlement would occur under the placement of 3–4 feet (0.91–1.22 m) of granular fill (Lea and Brawner 1963). In this instance, sawdust was placed to a depth of 5–6 feet (1.52–1.83 m) on the peat and the granular fill placed on top. Figure 6.10 illustrates the procedure employed.

The technique of preconsolidation has been used on many projects in Canada. Table 6.2 summarizes the published results for some of these projects. Preconsolidation has become widely accepted and has specific application where the depth of the peat is in excess of 6–8 feet (1.83–2.44 m) and where the final roadway grade need only be about 2–4 feet (0.61–1.22 m) above the natural ground line. Some differential settlement is to be anticipated. As a result, a flexible-type asphaltic concrete surface is preferable to a rigid Portland cement concrete pavement.

The first stage in the design requires that a detailed subsurface investigation be performed. Detailed testing is required to determine the physical properties of peat with specific emphasis on *in situ* shear strength and consolidation characteristics. The field and laboratory testing programme should be under the direct supervision of a qualified soils engineer.

The first actual step in the design is to establish the final desired grade and to designate the approximate location of any culverts or structures. The cost and time for construction will be minimized if the grade is established no more than 2–3 feet (0.61–0.91 m) above the surface of the muskeg.

A detailed Benkelman beam testing programme on existing highways constructed on peat has indicated that a minimum pavement thickness (gravel on

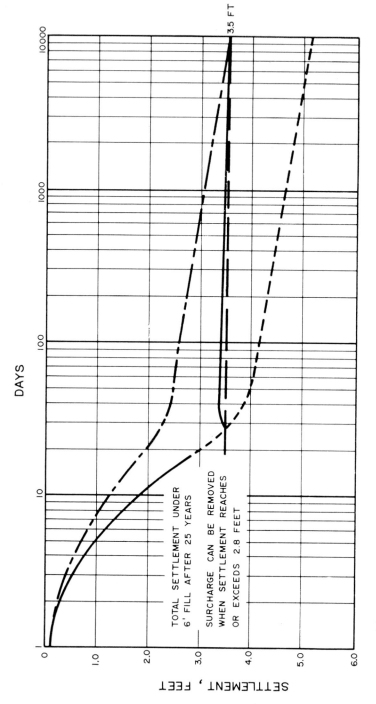

FIGURE 6.9 Time-settlement curves illustrating preconsolidation principle. — · — · —, time—settlement curve for 6-ft fill; – – – –, time—settlement curve for 11-ft fill; ———, time—settlement curve for 11-ft fill placed for 28 days, then 5-ft surcharge removed

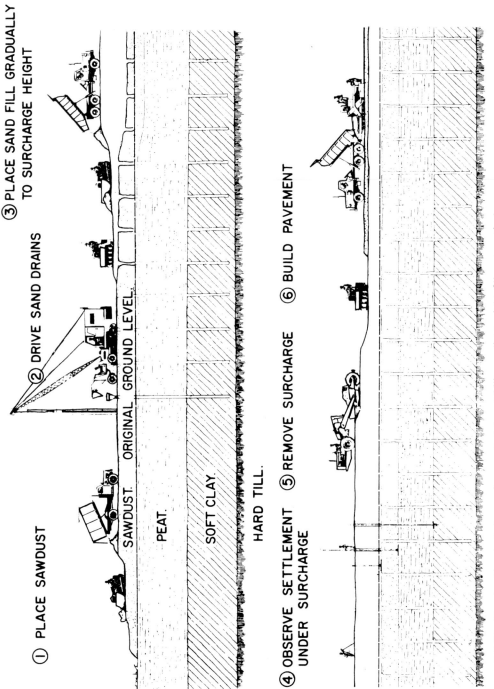

FIGURE 6.10 Construction technique using sawdust for stability as lightweight fill and vertical sand drains to increase the rate of consolidation of clay on the Burnaby Freeway, Vancouver, BC

TABLE 6.2
Summary of experience with preconsolidation

Project	Lulu Island, BC	Maillardville, BC	Burnaby Freeway, Deer Lake, BC	Prince Albert, Sask.	Syracuse, NY	Red Deer, Alta.	Napierville, Que.	St. Elie d'Orford, Que.	Orsainville, Que.
Date of construction	1957	1958	1959	1960	1959	1961	1963	1963	1960
Peat depth (ft)	6–10	20	30–35	3–6	3–5	10	5–11	3–13	4–14
Peat type	Fine-fibrous	Fine-fibrous	Fibrous-amorphous	Fine-fibrous	Fine-fibrous	Fibrous-amorphous	Fibrous-amorphous	Fibrous-amorphous	Fibrous-amorphous
Underlying mineral soil	Silty clay	Clay	Silty clay	Sand	Sand and silt	Silty clay	Clay	Silty sand	Silt
Shear strength (p.s.f.)	148–258	100–425	100–320	300 (av.)	240–300	210–520	140–190	120–300	100–600
Type of shear test	Field vane	Field vane	Field vane	Field vane	Unconfined compression	Field vane	Field vane	Field vane	Field vane
Water content (%)	1200–1700	500–1500	200–1200	220–430	280–400	100–480	300–650	200–890	200–700
Void ratio	21–29	8–27	3–17	3–10	3–5.5		7–11	5–16	3–12
Total fill placed (ft)	8	9	5	4–9	12	12	8	8	9
Total settlement (ft)	3.0	3.2	8.0	1.0	0.45–0.9	3.4	4.4	3.2	6
Surcharge height (ft)	3	4	2	2–4	4	8	3	3	2
Time of surcharge (days)	260	55	300–500	250	110	Not given	400	80 plus	350
Reference	Brawner (1959)	Lea and Brawner (1963)	Lea and Brawner (1963)	Mickleborough (1961)	Goodman and Lee (1962)	Anderson and Haas (1962)	Tessier (1966)	Tessier (1966)	Brochu and Paré (1964)

asphaltic concrete) of 3.5 feet (1.07 m) is necessary to provide adequate pavement performance (Fig. 6.4). Experience has also shown that a minimum surcharge of at least 4 feet (1.22 m) is desirable. This surcharge will result in an additional 1–2 feet (0.30–0.61 m) of settlement; therefore, a minimum total thickness of fill required will be of the order of 7–8 feet (2.13–2.44 m). If granular material is not economically available, the surcharge need not be gravel.

Settlement of the fill can be estimated from consolidation test results or from water content – void ratio – coefficient of compressibility relations (see Chapter 4). If no testing has been carried out, it can be assumed that settlement will equal one-half to one-third the depth of the fill placed, with the total settlement not exceeding approximately one-half the original thickness of the peat. As has already been mentioned, in very soft peat the total settlement may exceed the depth of fill placed.

Where extensive use of preconsolidation is proposed, it is recommended that consolidation tests be carried out. The standard method of performing these tests and predicting settlement by the use of the normal procedures of soil mechanics has been used with some success for predicting settlement in peat. A simpler technique, however, has been found to be sufficiently accurate for practical purposes. This technique assumes that consolidation is basically primary and that Darcy's law applies. Settlement in peat is then estimated by direct proportionality between laboratory tests and field tests. Undisturbed samples of peat are subjected to the actual load which is to be applied in the field and the settlement is determined and expressed as a percentage of the original height under that load. The field settlement is assumed to be equal to the same percentage (Hillis and Brawner 1961).

The time rate of settlement in peat is difficult to forecast by standard methods because of the great changes in the field permeability of peat and the coefficients of volume compressibility and consolidation. On completed projects in British Columbia, it has been found that the time required for settlement varies approximately as the thickness of the layer to the power 1.5. This can be expressed by the relation

$$\frac{t_{field}}{t_{lab}} = \frac{(H_{field})^{1.5}}{(H_{lab})^{1.5}} \qquad \text{(Lea and Brawner 1963).} \qquad (1)$$

Figure 6.11 illustrates the procedure used to determine the surcharge load required for peat treatment by preconsolidation. Curve ABC is a computed field load – settlement curve on which the abscissa of each point gives the settlement that will occur 25 years after the application of the load given by the ordinate.

Curve DEF is a similar computed field load – settlement curve for a duration of load equal to the allowable construction period, in this case assumed to be 3 months. Point D indicates that, if a load of 0.30 ton (272 kg) is placed and maintained, it will cause a settlement of 2.5 feet (0.76 m) after 3 months. Curve GDB represents the load which must be added to maintain the required finished highway grade.

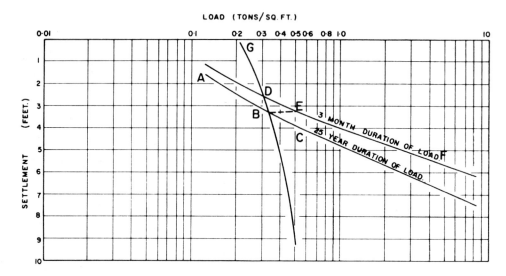

FIGURE 6.11   Chart illustrating procedure used to determine preconsolidation surcharge load

The slope and shape of this curve are influenced by the location of the water table and the unit weight of the fill materials. The horizontal projection of point B to line DEF gives the point E: the load required to achieve the total 25-year settlement in 3 months. In this example, the depth of fill to be added during pre-consolidation must then be selected so that, after 3.25 feet (0.99 m) of settlement, the load added to the peat will still be 0.52 ton per square foot (5075 kg per sq m). Design curves of this type must be constructed for each peat condition on a project.

If the settlement during construction differs from that indicated by the laboratory testing programme, the design chart must be modified in the light of the field experience. To determine fill quantities, the base width of the fill is increased 3 feet (0.91 m) for every foot (0.3 m) of settlement if 1.5 to 1 side slopes are used and 4 feet (1.22 m) for every foot of settlement for 2 to 1 slopes. After the required consolidation has occurred, the surcharge is removed to grade plus the depth equal to the crushed gravel requirement. If the construction over the muskeg is extensive, it may be feasible to use the excavated surcharge as fill further along the project.

The maximum height of fill that can be placed initially without shear failure can be estimated from

$$6\tau_f/\gamma, \tag{2}$$

where τ_f is the average shear strength of the peat and γ is the unit weight of the fill. Field vane shear tests have generally been found suitable for estimating the shear strength in the amorphous-granular and fine-fibrous peats (see Section 5.2 (1)). If soft clay underlies the peat, the shear strength in the peat–clay transition

FIGURE 6.12 Highway 401 under construction using preconsolidation. Instrumentation includes settlement gauges and piezometers

zone and in the upper layer of the clay must be determined. In evaluating stability, the different strain-to-failure rate of peat and clay must be considered.

Where extensive use of preconsolidation may be employed, more accurate analysis methods can be used to incorporate the gain in shear strength of the peat as the fill is placed and to consider the effect of porewater pressures. If fill in excess of that which can safely be placed is necessary, stage construction will be required. The maximum depth of fill is placed and the peat is allowed to consolidate and gain strength for 1–3 months before additional fill is placed. Field instrumentation to evaluate porewater pressure, settlement, and lateral movement should be designed and installed under the supervision of the soils engineer. Figure 6.12 shows a portion of Highway 401 near Vancouver instrumented to control construction. Figure 6.13 shows details of the stratigraphy, instrumentation, and settlement of a control test section for Highway 401.

Following the staking of the centre line, the field instrumentation is normally installed. Such instrumentation frequently includes settlement plates and piezometers every 100–400 feet (30.5–122 m). This instrumentation may be combined in one unit (Fig. 6.14). The piezometers should be installed near the shoulders. Lateral movement gauges placed at the toe of the fill are recommended to forewarn of impending lateral instability. These can be wood posts every 50–100 feet (15.25–30.5 m) driven on line about 15 feet (4.57 m) out from the toe of the fill.

After the field instrumentation has been installed and intial readings taken, the contractor may commence construction. Specific care should be taken to ensure that the surface mattress of the muskeg is not broken. Temporary drainage control is required at this time and culverts for this purpose can be placed on top of the peat. Initially, at least 3 feet (0.91 m) of fill is required in one lift to act as a working mattress for the equipment. This lift should be compacted by several passes of a vibratory roller. After each additional 1–2 feet (0.30 to 0.61 m) of fill has

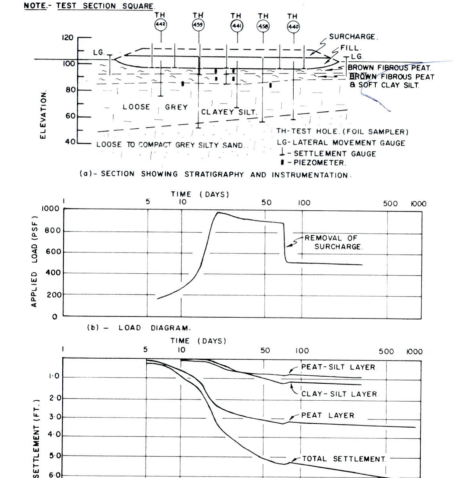

FIGURE 6.13   Details of stratigraphy, instrumentation, and settlement of a control test section for Highway 401 near Vancouver, BC

been placed, the instrumentation should be read and the data recorded and analysed. As a general guide, the contractor can be allowed to continue placing the material until pore pressures reach about 30 per cent of the total effective pressure applied or lateral movement exceeds 3 inches (7.62 cm). To facilitate later removal, material that is placed as surcharge should not be compacted.

After the grade has been completed to full surcharge height, field settlement readings must be continued to determine when the surcharge can be removed. Culverts are then installed to the desired grade by excavating trenches across the roadway. This is followed by placing of the base gravel and final paving.

(5)   Special Procedures

A   Vertical Sand Drains

Vertical sand drains have frequently been used to increase the rate of consolidation of soft compressible inorganic soils (Porter 1939; Moran *et al.* 1958). These drains consist of holes drilled or driven into the foundation soil and backfilled with clean sand. Their spacing and length depend upon the thickness of the compressible layer. The spacing of the drains varies from about 6–15 feet (1.83–

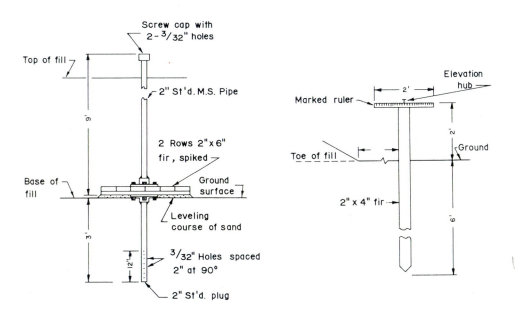

Settlement and Piezometer Gauge   **Lateral Movement Gauge**

FIGURE 6.14   Typical construction control gauges. Where deep piezometers are required they are installed through the M.S. Pipe depth. Normally ½–1 in. diameter pipe connected to a standard well point is placed in a predrilled 4-in. diameter hole with sand placed around the well point, and a bentonite seal placed above

4.57 m). A drainage blanket is placed on top of the natural ground, followed by the fill. The purpose of the drains is to reduce the length of the drainage path for porewater squeezed out during the process of consolidation. This allows more rapid dissipation of pore pressures and thus increases the rate of settlement and gain in strength.

In peat, secondary compression, which is not associated with pore pressure change, constitutes a moderate portion of the total settlement. Furthermore, the rate of drainage following the application of only minor loads becomes very low. As a result, it has been suggested that vertical sand drains are unlikely to have a beneficial effect on increasing the rate of settlement in peat (Hillis and Brawner 1961). Brawner (1958), reporting on field tests near Vancouver, noted that vertical sand drains did not increase the rate of settlement. The field observations revealed, however, that the sand-drain area was more stable and subject to less vibration than the portion of the road which was constructed as a floating fill without vertical sand drains. It was suggested that the sand drains tended to act as piles carrying the load to the firm underlying strata.

A summary of published evidence on the effectiveness of vertical sand drains suggests that their success in peat is questionable (Moran *et al.* 1958). If peat overlies soft compressible silts or clays, however, some advantage may result in the inorganic stratum.

B   Pile Foundations

Pile foundations have occasionally been used in muskeg for road construction. They are driven to firm material or to a depth to obtain sufficient friction for support. The piles carry a roadway slab, usually of reinforced concrete. Creosoted timber piles, steel piles, and concrete piles have all been used to support the slab. This method is, however, expensive and is limited for the most part to roads in heavily built-up areas where other methods of construction are not practical. Where steel or concrete piles are used, the problem of their corrosion and deterioration must be considered.

C   Chemical Stabilization

Some success has been achieved in stabilizing mineral soils by using materials such as resin, cement, chemical hardening agents, or bitumins. None of these materials has been found to be successful in stabilizing peat.

6.6   CONSTRUCTION COSTS

The cost of constructing highways through muskeg varies greatly, depending on the location of the project, characteristics, depth and extent of the peat, method of construction, source and type of fill material, drainage requirements, etc.

Construction costs summarized in Table 6.3 represent 1967 prices for average conditions in Canada. They are based on projects on major highways located in areas that are not isolated. A pavement width of 24 feet (7.32 m) and shoulder width of 10 feet (3.05 m) are assumed. The construction costs given in Table 6.3 should be increased by 50–100 per cent for projects in remote areas.

TABLE 6.3
Summary of construction costs in muskeg for average conditions in Canada (1967)

Item	Unit	Cost range ($)
Clearing		0–400.00
Corduroy mat (single layer)	Lineal foot	0.75–1.50
(double layer)	Lineal foot	1.25–2.25
Sawdust fill	Cubic yard	0.60–5.00
Dragline excavation		
(single throw)	Cubic yard	0.40–0.65
(double throw)	Cubic yard	0.75–1.10
Gravity displacement including granular fill hauled one mile	Cubic yard of fill	0.75–1.25
Jetting	Cubic yard of peat	0.25–0.40
Displacement by blasting including fill hauled one mile	Cubic yard of fill	1.10–1.50
Granular fill hauled one mile	Cubic yard	0.60–1.00
Granular fill pumped one mile	Cubic yard	0.50–0.85
Preconsolidation fill including surcharge removal – one mile haul	Cubic yard	0.80–1.30
Preconsolidation instrumentation	Lineal foot of highway	0.25–0.75

6.7   DRAINAGE

The location and design of drainage to control surface run-off in muskeg depend on whether the surface mat supplies some strength under the fill, on whether major settlement is anticipated, and on the method of road construction.

If the road is floated on the surface of the muskeg and the surface mattress provides some support, drainage ditches parallel to the road should be at least 50–100 feet (15.25–30.5 m) from the road shoulder to ensure that the mattress is not broken. Cross-drainage should be installed perpendicular to the centre line. Under no circumstances should culverts be constructed on piles since settlement of the adjacent fill may provide a non-uniform riding surface. Figure 6.15 shows a road constructed on muskeg with culverts on piles. The fill on both sides of the culvert settled about 18 inches (45.6 cm). To improve the riding qualities, the cover over the culvert was reduced and repaved.

Where large-scale settlement of the roadway takes place, for example when the preconsolidation method of construction is used, a natural depression occurs at the contact of the peat and the toe of the fill. This location can be developed into

FIGURE 6.15   Road floated on muskeg with culverts on piles. The patch on the highway is where the asphalt has been cut to reduce the bump above the culvert

a natural intercepting drainage ditch parallel to the centre line of the road. Because of the extensive settlement, temporary culverts are almost always required during the initial construction stage. Where complete excavation of the peat and replacement with stable material is employed, drainage procedures used for standard highway construction are recommended.

The selection of culvert locations and drainage ditch locations will generally be self-evident if a topographic plan of the area has been obtained. Natural drainage courses are inferred at low elevations and should be incorporated in the drainage design wherever possible.

Possibly the most significant decision which must be made is the selection of the type of culvert. In this connection, the following factors should be considered.

(1)   Settlement

No culvert will function satisfactorily if excessive settlement occurs. Consequently, efforts must be made to minimize this possibility. Metal culverts are best suited to resist structural damage when the road settles. Settlement up to about 2 feet (0.61 m) in 25 feet (7.63 m) is usually not serious enough to cause structural failure. Wood-stave culverts will withstand some bending. Concrete culverts should not be used where moderate settlement is expected.

(2) Lateral Movement

Significant settlement is also frequently associated with extensive lateral movement. Metal and wood-stave pipes offer reasonable resistance to this movement provided it does not exceed about 6 inches (15.25 cm) per 50 feet (15.25 m) of pipe. Concrete pipe should not be used unless it is placed upon a continuous reinforced cradle.

(3) Location

Location of the project relative to the source of supply and the method of transport available may have considerable bearing on the cost. In northern areas, wood stave pipe and aluminum pipe may be the most economical because of their light weight. To minimize transportation charges for metal pipe, several sizes of pipes may be included on the same order so that the smaller culverts can be nested inside the larger ones. Concrete culverts are generally uneconomical where long haul distances are involved.

(4) Installation

The costs of installation of steel, aluminum, and wood-stave pipe are reasonably comparable. The latter has the advantage of not requiring construction equipment to lift the sections into place. Concrete pipe may be very awkward to handle because of its weight and the special joint mortar that is required.

Installation techniques will depend on the road construction method. If the peat is completely removed and replaced or displaced with granular soils, normal culvert installation procedures may be used. Where preconsolidation is employed, several methods of installation are available. Flow across the grade must be maintained. During construction, the culvert may be placed temporarily and allowed to settle with the grade. Following settlement, it is dug up and installed at the desired grade. If the settlement is very great, it may be more economical to abandon it and place a new culvert. Occasionally, it may be beneficial to temporarily divert the channel until the required crossing can be stabilized. If there is little or no flow across the location, it may be advantageous to construct the grade, leave it to settle, and install the culvert later. Where settlement is not expected to exceed 1–2 feet (0.30–0.61 m), the culvert may be placed above the grade and allowed to settle gradually to the required grade. To allow for error, it is suggested that a culvert size be selected which has a capacity about 25 per cent greater than that actually required.

It is desirable that granular material for bedding be placed on the muskeg to a minimum depth of 12 inches (30.5 cm) and the width of three culvert diameters. The backfill around the culvert should be well compacted. This is particularly essential for wood-stave pipe. The majority of failures with this type of pipe have occurred as a result of faulty installation.

Culverts should, wherever possible, be placed perpendicular to the centre line of the road in muskeg. To provide maximum hydraulic efficiency it is desirable that an efficient inlet and outlet be utilized. Special flared metal entrances, concrete head walls, sodding, rip rap, or sand bags are frequently used. Where concrete culverts are installed, the joint mortar should be made with sulphate-resistant cement or high alumina cement.

In northern areas, icing of culverts and culvert entrances is a frequent problem. Culverts may freeze or ice build up in layers in the drainage channel and create mounds of ice which often exceed heights of 3–6 feet (0.91–1.83 m). Ponding areas for ice accumulation above the road, construction of burlap, tar paper, or wood fences along the road, or using two culverts, one higher than the other, may be effective in controlling this ice build-up.

(5)   Corrosion or Deterioration

Steel culverts are frequently subject to severe corrosion in muskeg waters. This is due mainly to the relatively high quantities of soluble salts and high total acidity which frequently occur in muskeg waters. Romanoff (1957) presents a correlation between soil resistivity readings and the amount of soluble salts in the soil moisture. Readings taken in the soil in the vicinity of the culvert provide a method of estimating if the water is likely to be corrosive. Several methods of protecting steel culverts are available and are discussed in detail in Section 7.7(2).

Aluminum culverts have been used to a limited extent in muskeg. Godard (1955) reports some success with various aluminum alloys buried in highly acidic soils for periods up to 5–10 years. Long-term tests, however, have not been carried out.

Concrete is subject to attack from soluble salts, especially sulphates and chlorides. In addition, the decomposition of organic matter produces carbonic acid which may be detrimental to the concrete. The best way to minimize deterioration of the concrete is to use a dense mix and entrainment of 6–8 per cent air. Even with good concrete, however, $\frac{1}{2}$–1 inch (1.27–2.54 cm) of deterioration can be expected over a long period of time. Extra thickness of concrete should therefore be specified if concrete culvert pipe is used.

Creosoted wood stave culverts have been used very successfully in British Columbia in muskeg areas. Provided the pipe has been properly treated, 20–25 years of service can be expected from this type of culvert. The metal bands may corrode rapidly, but since they are used basically to stress the pipe during installation, this is not a serious problem.

6.8   MAINTENANCE OF HIGHWAYS CONSTRUCTED IN PEAT AREAS

The major problem associated with the maintenance of roads constructed on peat is that of maintaining a uniform grade. Differential settlement frequently occurs and the least expensive method of improving the riding quality is to completely fill all depressions. This extra weight, however, often produces further settlement and a follow-up sequence of fill maintenance will be required at a later date. An alternative, but not so common procedure, is to remove the humps in the road. This is initially more costly, but no extra load is imposed on the peat and future settlement is minimized.

On roads that are floated over the muskeg, it is quite common for lateral movement, resulting in tension cracks parallel to the centre line of the road, to take place. These cracks should be filled with a slow-curing asphalt. Many applications may be required before the movement ceases.

If extensive alligator cracking develops in the road surface, this indicates that the thickness of the pavement structure comprising the asphalt and sand or gravel base and subbase is not sufficiently thick to reduce the fatigue stresses in the asphalt pavement to tolerable limits. Additional thicknesses of granular material and asphalt will usually be required in this case. The additional thickness required may be determined by using the Benkelman beam rebound test (Brawner 1963).

A general inspection and maintenance programme for culverts, particularly for those in muskeg, is desirable. Not only will such a programme provide greater assurance of proper drainage control but it will also provide valuable data for assessing the success and application of various types of culverts for various conditions.

PART II: SECONDARY AND ACCESS ROADS

6.9   INTRODUCTION

A secondary road is one that does not have the importance of a prime arterial traffic route. It could and may develop into one at some future time when an increase in population or development of industry demands it. One must, therefore, keep in mind the possible salvage of work already undertaken. Access roads can very easily be grouped with secondary roads since they may be secondary in nature to the overall road network plan. Possible salvage must be kept in mind where comparatively large sums of money are being expended.

This part of the chapter will suggest the treatment of construction for those roads that are not primarily in the major highway set-up. These may be township roads, minor county roads, or access roads to industrial developments built for use during the construction of an industrial or utility complex such as a hydro-electric project. Such roads may require intensive use for periods of up to 2–3 years. Forestry access must also be considered both for the prevention and fighting of fires and to serve parks and recreational areas. The intensity of treatment should closely parallel the importance of the end result. Expensive construction required for a major project would certainly be a waste of money for an "in and out and gone" effort which might be of little use in the future.

As the search for natural resources moves farther and farther afield, we come increasingly face to face with that vast area of the north country where muskeg forms the greater percentage of the area. Ideas of road building conceived for southern regions need to be tempered by and augmented with the requirements for a different type of terrain, where solid rock is interspersed with increasing areas of organic terrain and weaker, finer types of inorganic soils. Often peat is preferable to the siltier soils since one of the constituents of the former may be a woody network of fibre, which may offer greater stability for road foundation.

Secondary and access roads, when built in accordance with the proper method, have proved successful throughout the country. Careful summarization of the data of available case histories and adaptation to a pertinent need will aid both the design and the construction engineer and will help to minimize the costs of construction. Careful documentation of the results achieved and further research programmes covering the "unknowns" will help to standardize procedures and prevent past errors.

6.10   LOCATION AND GEOGRAPHICAL CONSIDERATIONS

Location methods employed for secondary roads closely follow the pattern used for more important arteries, except that the detail may not be required. Photogrammetric methods are used initially. With the Radforth approach to aerial interpretation of muskeg (see Chapter 3) great advances have been made in avoiding muskeg altogether or in selecting routes that offer better structural types of organic terrain.

Initially, a route is chosen on 8 mile-to-the-inch topographic maps after an aerial reconnaissance by helicopter has yielded the general characteristics of the area. Basically, this route should be the most direct, with due regard for the fundamental obstacles to be avoided in its location. Airphoto cover sufficiently wide to take care of alternative routes is then obtained and a photogrammetric location projected thereon. Muskeg classification is also derived from the photos by using the Radforth technique. A field check is made and the projection, with some revision during the field check, is blazed on the ground. Subsequently, a survey line is cut and the survey and plotting carried out. The field check will generally confirm assumptions made from the photos with regard to classification, and the revisions will be the normal refinements made in this type of location. During the survey, soil and vegetation are typed and the muskeg is probed to determine the depth and character of the peat.

6.11   PRELIMINARY STUDIES

The breadth and depth of the studies carried out for secondary and access roads should follow closely the pattern used for primary highways, except that care must be taken to keep in mind the relative importance of the road. This is equally true when dealing with design.

Preliminary evaluations leading to the actual design should cover foundation soil characteristics, depth and type of organic terrain traversed, striking of the grade line (profile), availability of borrow and of granular and surface materials, type and placement of drainage structures and ditching (offtake and lateral), and incidence and type of traffic. To bring the complete picture into proper focus, studies carried out in the office must be closely related to field investigation, and to the data derived therefrom.

6.12   DESIGN CRITERIA

Generally, the following standards will produce a road adequate to support moderate to heavy movement of light traffic with the occasional transport of heavily loaded vehicles (up to 40 tons (44.08 tonnes) total weight). The completed road will afford all-season travel with ordinary maintenance.

(1)   *Road Width*
The profile grade at the top of the surface should be as follows:
   Single lane – 16 feet (4.9 m) between shoulders,
   Two-lane – minimum of 24 feet (7.3 m) between shoulders.

Usually a single-lane road of the above width with drive-outs at strategic intervals to accommodate passing vehicles will be adequate. The terrain will influence the width of the section. In hilly or mountainous country drive-outs will be required more often, the design width being dictated by the sight distance. In cases where the incidence of traffic will be high, a two-lane standard is obligatory. When dealing with muskeg, it is safest to design the subgrade as a two-lane section and economize on top granular and/or on surface courses in width, i.e., a two-lane subgrade with one-lane granular and surface courses. It is difficult to reconstruct and widen subgrades in muskeg areas. This is particularly true where corduroy or brush matting is used to carry the embankment over the organic terrain since differential settlement may result.

The incidence of traffic and the type of traffic (passenger vehicle, bus, truck, tractor, float, etc.) must be assessed and the requirements as related to the end result evaluated.

(2)   *Height of Embankment*
It has been established that a compacted embankment 3 feet (0.91 m) higher than the level of the adjacent muskeg is adequate.

(3)   *Depth of Granular Base and Surface Courses*
Over a subgrade composed of suitable material, 12 inches (30.5 cm) of selected granular base course Class B and 6 inches (15.25 cm) of Class A surface

material should be adequate. Greater depths of granular base course will be required over common material which does not meet the specifications or when placed over borderline material under unfavourable weather conditions.

(4)   *Grading Requirements for Granular Base and Surface Courses*
The two classes of granular material should conform to the grading requirements of Table 6.4 which shows the percentage passing the various sieves. Class A material should be produced by crushing pit-run gravel or stone. Class B material may be used directly from the pit without processing if the *in situ* material conforms to the specification requirements. Screening and/or blending with material from other approved sources may be required to produce the specified grading.

TABLE 6.4
Grading requirements for granular base

Sieve size (square openings)	Percentage passing	
	Class A (%)	Class B (%)
4 in.	—	100
$2\frac{1}{2}$ in.	—	—
$\frac{7}{8}$ in.	100	57–100
$\frac{5}{8}$ in.	75–100	—
No. 4	35–60	25–100
No. 14	15–35	10–85
No. 48	5–20	5–40
No. 200	3–8	3–8

(5)   *Drainage Structures*
Because of acidity, culverts specified for placement in organic terrain should be of protectively coated metal pipe or treated timber, as pointed out in Section 6.7(5). Corrugated metal pipe with asphaltic coating and round or rectangular creosoted timber types have given excellent service.

6.13   CONSTRUCTION METHODS

Depending on the muskeg conditions, three methods of construction are considered:
   Peat depth under 3 feet (0.91 m) – excavate,
   Peat depth under 3 feet – ride,
   Peat depth over 3 feet – corduroy.
   Embankments in muskeg over 3 feet (0.91 m) in depth without corduroy and without excavation are fully discussed in Part I of this chapter.

(1)   *Clearing*
One hundred feet (30.5 m) is specified to facilitate working room and main-

tenance. This width will also afford non-interference with the muskeg within a reasonable distance of the roadbed. Trees and vegetation are usually cut at heights from 18 to 24 inches (45.7 to 61 cm) above ground to make grubbing operations easier.

(2)   *Close Cutting and Grubbing*

When the peat is left in place and the embankment is either "floated" directly on the muskeg surface or supported by corduroy, close cutting is called for and the natural mat of the muskeg must not be disturbed. In these sections every care must be taken not to fracture this mat and no vehicles should be permitted inside the 100 foot (30.5 m) right-of-way until the corduroy is laid. Grubbing is usually carried out as part of the operation if the peat is excavated.

(3)   *Placing of Embankment Material*

Based on current costs, the point of economic balance between the cost of the grubbing and excavation of peat and subsequent backfilling and the cost of the supply and placement of corduroy falls at about a peat depth of 3 feet (0.91 m). For depths less than this it is considered advisable to excavate for the full width of embankment. The excavated peat is deposited outside these lines and later, after the embankment is completed, brought back to flatten the slopes, thus deterring and possibly preventing washing and gullying.

For peat depths of less than 3 feet (0.91 m), where it is known that the structural strength of the underlying mineral soil is inferior to that of the peat, it is advisable to ride the close-cut surface of the muskeg with the embankment. Placement of the fill material can best be accomplished by end-dumping. In order to preserve the natural fibre of the mat construction, equipment should not be permitted to traverse the muskeg.

(4)   *Corduroy Construction*

Where deeper muskegs are encountered and corduroy is placed, the initial fill over the corduroy mat should be 12 inches (30.5 cm) in depth, suitably compacted. Succeeding layers should be compacted to achieve 6-inch (15.25-cm) lifts of compacted material.

Corduroy may be constructed by either of two methods:

(a) Figures 6.16 and 6.17 show a high standard of construction. Longitudinal stringers are spiked with half-inch (1.27 cm) drift pins or wired to each log. All logs must have a minimum tip diameter of 6 inches (15.25 cm). Joints in the stringers and courses are staggered. Branches and smaller trees are placed on top of the logs to chink voids and to prevent the embankment material from passing through to the muskeg and creating voids leading to the development of pot-holes. Coniferous trees give the best results.

(b) A cheaper form of construction (used in less important roads) is carried out by laying the cleared trees side by side (without limbing) across the centre line of the road. Smaller trees and brush are placed on top to make a dense, compact mass which will flatten down to some extent beneath the embankment. The ultimate depth of the mat should be of the order of 18 inches (45.7 cm).

In both cases the mat should be completely covered by the embankment material so that air is excluded. If these instructions are followed, the mat should remain sound indefinitely.

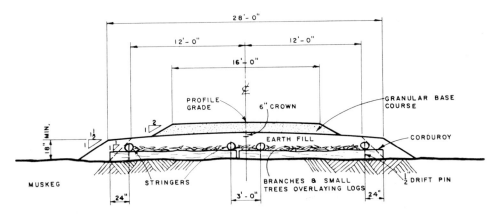

FIGURE 6.16   Earth fill on corduroy

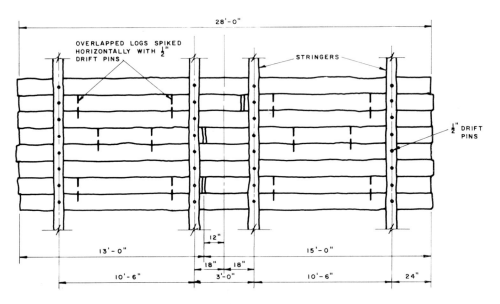

FIGURE 6.17   Plan of corduroy

6.14 DRAINAGE

Ditching in organic terrain should be organized with the least possible disturbance to the natural drainage pattern. Lateral ditches may be required to maintain the natural water table. Where necessary, they should be far enough from the roadbed to eliminate the danger of a fracture in the embankment. Experience indicates a distance of 20 feet (6.1 m) from the toe of the embankment to the nearest edge of the ditch.

Culverts should be placed on inorganic soil at the ends of muskeg areas, wherever possible. This will minimize settlement by providing an adequate foundation. Where equalizers are required it will be necessary to bed the culverts on a sufficient depth of compacted granular fill to maintain the correct invert elevation. In exceptional cases it may be necessary to build rough crib foundations. In deep muskegs where corduroy is used the culverts should be placed on the corduroy mat. In all cases the invert should be set to maintain the existing water regime since lowering it will cause deterioration in the structural strength of the natural muskeg mat.

6.15 MAINTENANCE

The maintenance of properly constructed secondary roads on organic terrain should pose no greater problems than that of roads constructed on mineral soil. Regular inspection, however, with certain features in mind will pay dividends. These features include:

1. Development of irregularities in the grade line or in the cross-section line indicating settlement, either general or in pockets. The former may indicate movement in the peat underlying the embankment, where corduroy has been placed, or in the underlying mineral soil where excavation of peat has been carried out. Pockets of settlement are usually the result of voids developing in the corduroy mat, possibly due to faulty construction. If the settlement is minor, filling with appropriate material is indicated; if serious, continued inspection may indicate that a general regrading and reconfirmation of the road section are required to restore the grade line and cross section.

2. As mentioned in Section 6.14, the natural water regime should be intact after construction. Vigilance is required here. Radical changes in the water table may be caused by faulty culverts and impaired ditching. These faults should be corrected immediately. A lowered water table could result in deterioration of the muskeg and consequent settlement of the road. A higher water table could endanger the embankment itself.

3. Lateral displacement of the embankment, caused perhaps by movement in the peat itself, may develop. This is the most serious condition that could occur

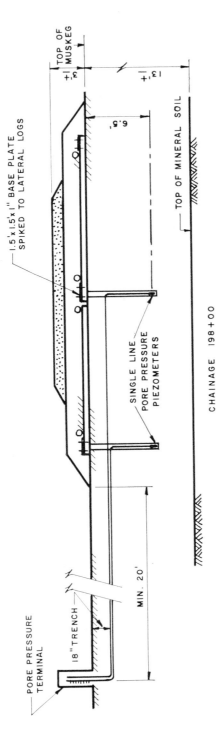

FIGURE 6.18 Instrumentation on muskeg section, Little Long access road, Northern Ontario

TABLE 6.5
Little Long–Pinard road muskeg area instrumentation

Station 198+00

	Settlement plates			Piezometer readings			
Date	10 ft left of centre line Original elev. 686.41	Centre line Original elev. 686.34	10 ft right of centre line Original elev. 686.48	#1–centre line Original reading 0.0	#2–13 ft left of centre line Original reading 0.0	On corduroy	Remarks
1962							
July 11	686.41 (0.0)*	686.34 (0.0)	686.42 (0.0)	0.38 (CL)†	0.36 (CL)		No fill
July 11	686.28 (0.3)	686.22 (0.5)	686.42 (0.2)	3.00	1.01		Fill being placed
July 12	685.94 (1.4)	685.71 (1.4)	685.89 (1.3)	3.48	0.47		⎰ Fill being placed
July 17	685.12 (2.2)	685.26 (2.5)	685.44 (2.0)	4.12	BP†		⎱ and compacted
July 27	684.70 (2.6)	684.35 (2.8)	685.25 (1.5)	4.20	BP		
Aug. 23	684.68 (2.6)	684.85 (2.8)	685.39 (1.5)	4.74	BP		
Sept. 22	684.65 (2.4)	684.83 (2.6)	685.29 (1.7)	4.71	BP		
Oct. 26	684.60 (2.5)	684.75 (2.9)	685.23 (1.9)	5.34	BP		Granular base being added
Nov. 9	684.47 (3.1)	684.62 (3.5)	685.12 (2.6)	4.90	0.90		
1963							
Jan. 3	684.34 (3.3)	684.55 (3.6)	685.00 (2.6)	4.80	BP		
Apr. 18	684.28 (3.4)	684.60 (3.5)	685.00 (2.5)	5.14	1.33		Grading operation
May 17	684.24 (3.4)	684.56 (3.5)	684.97 (2.5)	4.96	1.08		
July	684.14	—	684.70	4.78	1.22		Final lift of granular base
1964							
Jan. 24	684.10	684.45 (3.7)	684.61	3.70	BP		
1965							
June 21	Not available			3.67	0.96		

*Readings in parentheses are depth of fill.
†CL, corduroy laid; BP, below platform.

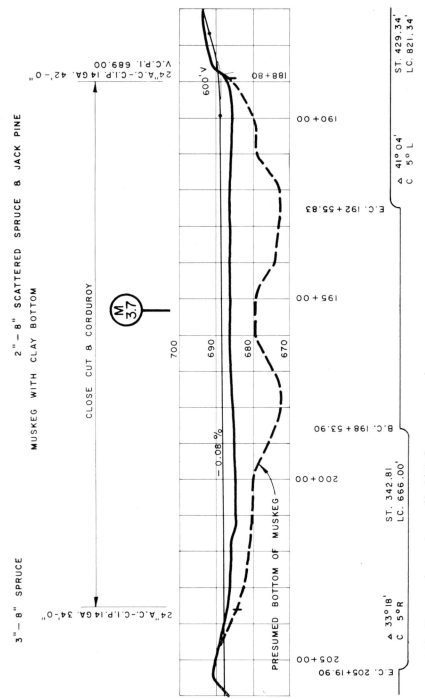

FIGURE 6.19 Deep muskeg section profile, Little Long access road

since it is the most costly to remedy. Investigation would be required to determine the cause of the shift and the possibility of constructing a berm on the downstream side to stabilize the movement.

6.16 A CASE HISTORY

Unfortunately, much knowledge is lost because of the lack of documentation of specific construction procedures on various projects. Since secondary and access roads (important as they may be) occupy a minor position in the general picture of road construction, they are all too often neglected in the general scheme, particularly with regard to the expense of instrumentation and documentation.

On one such access road, however, constructed during 1962–3 in the Little Long area of the Moose River basin of Northern Ontario, a careful record was compiled of the method of construction. Instrumentation was carried out (see Fig. 6.18) and the results tabulated. Settlement plates were installed at the centre line and under each shoulder of the road on the corduroy at the bottom of the embankment material. Piezometers were also installed at a depth of 6.5 feet (1.98 m) below the top of the muskeg, beneath the centre line and the right shoulder, with the terminal returns to staff gauges above the surface at a distance of some 20 feet (6.1 m) from the toe of the embankment slope.

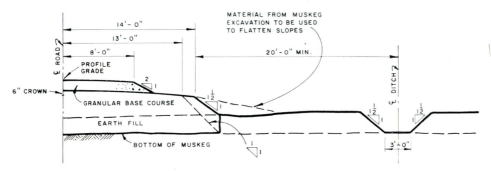

FIGURE 6.20   Muskeg excavation, Little Long access road

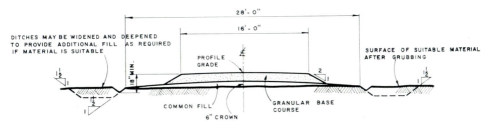

FIGURE 6.21   Graderwork section, Little Long access road

Abbreviated results of readings taken at various times are shown in Table 6.5. This partly illustrates the behaviour of the embankment during and after this particular construction. A profile showing the depth of the muskeg and the grade line to which the embankment was constructed is shown in Figure 6.19. Typical sections of the road are shown in Figures 6.20 and 6.21. Little or no distortion has occurred in this section of road since it was constructed. It has consistently supported loads in excess of 40 tons (44.08 tonnes) gross over protracted periods of time with no apparent ill effects except understandable wear to the surface which was composed of 6 inches (15.2 cm) of Class A crushed rock.

PART III: RAILWAYS

6.17 INTRODUCTION

Tramways consisting of cross timbers and rail made of oak were used by horse-drawn coal wagons in Britain in the seventeenth century. With the introduction of iron rails, this system of track was found to be so strong, adaptable, and flexible that it was used for train traffic and has continued in principle up to the present time.

In specifying construction requirements for track it is important to remember that a track structure can easily be lifted and levelled periodically and even shimmed against temporary unevenness of support. While it is obviously desirable to minimize such maintenance operations, construction standards for railway track are not as exacting as those for a paved surface.

On the other hand, railways cannot avoid crossing soft ground as easily as roads, since the flat grades frequently require that track follow river valleys and forbid climbing through hilly areas which may surround boggy ground. It is important also to avoid excessive curvature in order to maintain speed and minimize hauling power and wear on rail and wheels. The values assigned to various aspects of location are therefore somewhat different for railways than for highways (Monaghan 1963).

The traditional method of building track through muskeg areas which could not be avoided was simply to keep the grade low and try to float the fill. If failure occurred, fill was added repeatedly until the grade stayed in place. This method was used, often with corduroy logs, in building all the pioneering railways in Canada. The many stories told of locomotives and cars being involuntarily included in fills on muskegs are by no means all imaginary.

Since maximum axle loads have not increased since the days of the early large steam locomotives, most of these muskeg crossings are still serving satisfactorily. Because railroaders are a taciturn group, however, the high costs that must have been involved in maintaining newly constructed lines are not recorded – and indeed have been taken for granted until the present generation.

With the evolution of modern earth-moving equipment, railway construction procedures in muskeg areas have become more flexible. Most of the techniques used in highway construction have been adapted to new railway construction. In addition, there has been considerable improvement work carried out on existing muskeg crossings. This has often required special procedures to keep the line open to traffic during the construction period.

6.18 PLANNING CONSIDERATIONS

The concept of building roadbeds for a specific design life is not used in railway work. With upgrading of the track structure and improved signal systems, a single-track railway can normally handle all the traffic available except on the most heavily travelled routes. A railway grade is designed, therefore, to last indefinitely with periodic upgrading. Figure 6.22 shows a mainline railway in a muskeg area in New Brunswick which was built in 1910. This track is in excellent condition. Once initial stability is achieved in a muskeg area it is usually affected only by an increased load applied on the foundation material.

Basic to planning, of course, are considerations of the type of traffic to be handled, axle loads, frequency, and speeds, on which standards of construction are chosen.

(1) Alignment and Grade Requirements

Although it may seem paradoxical, the higher the standard of construction chosen, the less chance there is of avoiding muskeg areas in locating a route. A first-class location involving tangent track, flat grades and curves has limited ability to pass around muskegs. Where muskegs are unavoidable, it is standard practice to keep all fills as low as possible.

(2) Time Available

Time is a major factor in most railway design and construction. Since new rail lines are normally built to serve industrial developments, construction time must be kept to a minimum, with a strict deadline for completion. The construction schedule is drawn up on the assumption of normal weather. Conditions worse than normal may require a change in the approach to some problems.

On the other hand, the most economical method of treatment is often the use of time to allow organic soils to consolidate and gain strength. When possible, stabilization periods of a few months, particularly during winter, should be incorporated into the construction grading schedule to reduce costs with a minimum of inconvenience to contractors.

Winter grading techniques are being developed, and are of particular value where borrow material is scarce since less material is wasted on access roads.

FIGURE 6.22   Main line for heavy traffic built in 1910 across muskeg in New Brunswick

FIGURE 6.23   A small dragline on pads is not safe on weak peat

FIGURE 6.24   Dredged sand fill being placed in a completed dragline excavation through weak peat. Sloughed peat is being displaced and pushed ahead by the fill. The resulting mixture of fill and peat will cause irregular settlements of the grade

(3)   *Scheduling and Access*

It is not sufficient to assume that exploration and construction will proceed in an orderly manner along a route from A to B. Priorities must be set. Difficulties of access for exploration and type of treatment required for muskeg areas suggest that work should be started at some points in advance of similar work elsewhere.

In muskeg areas, access for crews responsible for ground surveys and exploration must be planned carefully. The desirable amount of exploration before construction is a matter of judgment based on experience and the ease of access at different times of the year at different stages of construction should be taken into account. Over soft muskeg areas, access of drilling equipment may be impossible except on the frozen surface during the winter months.

(4) *Muskeg and Underlying Soil Conditions*
It is most important to decide on the method of building a grade across muskeg on the basis of prior information on depth and physical properties of the soil. Excessive costs may result if this is not done and a change in plans is required. It is therefore essential that muskegs be explored in some detail before construction reaches them. For an individual muskeg this can be done after the contract is let. For the project as a whole, sufficient information on the extent, depth, and characteristics of the most important muskeg areas and proposed methods of dealing with them should be available at the time of tendering.

(5) *Construction Materials*
The selected method of treatment will often depend on the availability and cost of backfill material. For example, a displacement fill may require several times as much backfill as a floating fill. The cost of fill material at the site should, therefore, be estimated in the preliminary design.

Replacement fill is normally supplied by trucks or scrapers and may be supplemented by train-hauled material after the track is in place. Dredged sand or gravel may be economical under particular conditions.

(6) *Adjacent Structures*
The question of possible damage to adjacent structures needs careful consideration in planning. It is discussed in Section 6.2(4) with reference to adjacent roads and approach grades for bridges. In railway maintenance there is the additional problem of replacing timber trestles at the end of their useful life without interruption to traffic.

(7) *Construction Problems and Control*
This subject is discussed in Section 6.2(6). Because of time limitations railway grade construction and control are sometimes more difficult than for roads. Advance information on site conditions can help the contractor to anticipate difficulties and plan how best to handle them. If a plan for each muskeg crossing is made in advance, the necessary controls as outlined in Section 6.5 can also be arranged in advance.

6.19   EVALUATION OF ECONOMICS

Remarks in Section 6.3 on the cost factors of right-of-way, design and engineering, safety potential, user benefits, and construction and maintenance may all be usefully applied in railway planning economics.

6.20   SITE EXPLORATION

Muskeg areas are one of the most easily recognizable landscape features in air-photos. For this reason airphotos are an important tool in preliminary railway route selection through muskeg territory (see Chapter 3 and Section 6.4). Airphoto interpretation can give some indication in advance of ground surveys of the severity of the engineering problems to be expected at individual muskeg locations.

Exploration must usually be staged to accommodate the scheduling of design and construction work. The stages suggested for the best co-ordination of effort are shown in Table 6.6. Stage 1 exploration is planned so that major problem areas

TABLE 6.6
Stages of exploration prior to construction of railways in muskeg areas

Stage	Purpose	Work
1	To find depth profile of muskeg and firm bottom; assist in appraisal of alternate routes	Hand soundings with steel rod, shear vane tests
2	To find extent, depth, and properties of muskeg; assist cost analyses in decision on final route; provide information for contract documents	Sample borings and vane shear tests correlated with additional soundings
3	During construction to provide detailed information required; "trouble shooting," development and appraisal of treatment techniques	Sample borings, shear vane tests, field measurements

can be located and evaluated with a view to judging the construction costs involved in alternative locations. Stages 2 and 3 sometimes overlap on small projects or where access is difficult on large projects. On large projects, stage 3 exploration continues until the grading work is completed.

Because of the degree of judgment required in interpreting and correlating site exploration, it is important to have an experienced engineer, assisted by technicians as required, in charge of this work.

6.21   DESIGN AND CONSTRUCTION

There is a tendency to gamble in advancing a line across muskeg to provide access in a remote area. Costs are increased by delays, whether due to difficulties which may develop if simple construction procedures are used, or due to adaptation of more complex procedures to avoid those difficulties. Preconstruction planning, exploration, and design will greatly reduce the probability of construction difficulties and hence change the odds.

The application of design to specific crossings is also important because methods of construction are not easily interchangeable once work has started. If a fill which is intended to float breaks through the surface mat, more and more material is added, thus weakening the muskeg by disturbance until the job turns into an attempt to displace the peat and get the fill down to a firm bottom. In this sequence of events, costs are much higher because of the excessive amounts of fill used, and results are much poorer because of peat pockets trapped under the fill and the disturbed muskeg containing it, than would have been the case if a displacement fill had been planned and built in the first place.

Muskeg behaviour under load is generally predictable. "Practical" people who claim otherwise are going against many of the considered opinions in this Handbook.

In all methods of construction across muskeg, initial grade width should be liberal to allow for future settlements, particularly in view of the maintenance difficulties to be expected if more fill is added later (see Section 6.23). At least 4 feet (1.2 m) of fill and preferably more should be placed over peat under all conditions to provide a working pad and reduce frost heaving. A discussion of drainage in construction operations is given in Section 6.22.

(1)   Floating Fills

This method is applicable to low fills when there is a good surface mat of vegetation underlain by relatively firm peat and when the deposit is too deep to excavate economically. It is commonly used with fills 4–5 feet (1.2–1.5 m) high on muskegs up to 15 feet (4.6 m) deep. The method is also recommended when soft silt or clay underlies peat. Excavation and replacement or displacement methods under such conditions have the effect of subjecting the soft silt or clay to a greatly increased load, which leads to substantial settlement or possibly failure. Construction on the surface of the muskeg minimizes this problem.

The advantage of this method is its low first cost. The disadvantages are serious differential settlement of fill, difficulty in avoiding displacement, and high maintenance costs. Detailed procedures for the design and construction of floating fills are given in Section 6.5(1).

Owing to the relative ease of maintaining track levels, floating fills are quite satisfactory for railway grade construction under many conditions. Safety against failure must be checked in advance of construction, however, to ensure that an impromptu, incomplete displacement of the fill, with its attendant complications, does not result. Drainage in advance of construction is important.

Winter construction of floating fills on muskeg has some advantages which should be noted. From limited but encouraging experience (Cronkhite 1967), the following procedure is suggested: (a) stripping right-of-way before winter to assist penetration of frost; (b) removing snow from roadbed area when sufficient frost has penetrated to support equipment; (c) advancing drainage ditching; (d) constructing embankments up to grade, placing fill in short sections so that normal layers are compacted and covered before serious freezing takes place; (e) applying compaction in the normal manner but with no attempt to adjust moisture content; (f) shaping the grade for quick surface drainage.

(2) Excavation and Replacement

This method involves excavation of peat and replacement with stable backfill material. It can be used best with shallow muskeg and plentiful low-cost stable backfill. It is not recommended for use next to structures because of the possible lateral movement induced.

The advantage of this method lies in the opportunity to obtain a stable roadbed with little settlement. Disadvantages are the cost of the large amount of backfill required and the possibility of differential settlement of fill if pockets of peat remain unexcavated or slide in from the excavated slopes.

In its simplest form this method involves the stripping of organic material from fill foundation areas. This is advisable when the peat lies within the depth of frost penetration under a fill because of the differential frost heaving which frequently occurs. Partial excavation of peat before fill placement is not recommended. Full excavation and replacement of peat between 4 and 12 feet (1.2 and 3.7 m) in depth should be decided after the costs and benefits have been compared with those of floating and displacement-type fills.

Construction methods are discussed in Section 6.5(2). As railway grades are not wide, it is usually not feasible to carry on both excavation and backfill operations from the end of the fill. Commonly, a dragline operates on pads on the muskeg in front of the fill, excavating and casting material to the sides. The allowable size of dragline depends on the bearing strength of the muskeg and miscalculations are expensive (see Fig. 6.23). Fill is usually dumped from trucks or scrapers and bulldozed into place. Excavation and fill operations are kept as close together as possible.

Occasionally it is possible to do either excavation or backfilling by dredging. Dredge excavation avoids the problem of spoil banks as the material can be pumped some distance away. A dredge, however, can dig or deposit material at a much higher rate than any land-based equipment. Thus, if the dredge is digging, the excavation is opened up far ahead of the fill operation. If the dredge is filling, the excavation must be nearly completed before filling can start if a continuous and economical dredge operation is to result. The latter type of operation in a soft muskeg is shown in Figure 6.24. Excavation slopes have sloughed in some places and peat is being pushed ahead of the fill and mixed with it. Excessive fill quantities and differential settlement of the finished grade may result from this type of operation.

The disposal of material should be planned in any excavation and replacement project. The plans should require that the waste material be cast far enough away to eliminate danger to the excavation slopes. The material may be used later to flatten the slopes of the completed fill. The considerable shrinkage of organic soils on drying (see Section 4.2(2)) should be kept in mind in planning final cross-sections.

(3)   Displacement Methods

The planned and controlled displacement of muskeg with fill material can be carried out by various standard methods. These are applicable where depths of muskeg are between 10 and 30 feet (3 and 9 m) and special techniques are said to be successful with depths up to 50 feet (15 m). The method is not recommended for use next to structures because of the lateral movement induced.

If reasonably complete displacement can be achieved, this method has the advantage of obtaining a stable fill with little settlement after construction. Its disadvantages are the likelihood of trapping pockets of peat when the displacement is not done carefully, causing differential settlements in the completed grade. Cracking and movement of the shoulders of the fill may also occur, requiring wider than normal grades to compensate. Fill may be any stable material, but coarse bank-run gravel or rock fill is preferred in order to obtain penetration of the surface mat of muskeg.

Success depends upon ensuring that enough embankment weight is added and that displaced material is excavated deep enough before the advancing fill front so that the direction of displacement is controlled. The advancing front should always have a steep face and the rate of filling should not be faster than the rate of removal of displaced material. It is particularly important to remove peat which heaves above the ground water level.

The effect of the slope of the firm bottom underlying muskeg is often overlooked in displacement operations. It is difficult to displace peat against a rising bottom. If the bottom is known to be higher on one side, the front of fill can be skewed to direct displaced peat towards the deeper area and the direction of successive fill fronts can be shown on construction plans.

On the other hand, when the slope of the firm bottom is falling away, an unexpectedly large amount of fill may be needed. An example of this occurred where a surcharged gravity displacement fill of sand was placed along the edge of a railway yard near Pontiac, Michigan, to prepare for additional tracks. The fill was across one side of a saucer-shaped depression in glacial till at least 50 feet (15 m) deep in the middle and containing about 15 feet (5 m) of very soft silt overlain by weak peat. A section through the area is shown in Figure 6.25. The limits of fill as planned can be compared with those which resulted. Before the surcharged fill reached stability, three times the estimated quantity of fill was required and the area appeared as shown in Figure 6.26. Despite a relief trench kept open in front of it, the fill material ran under the peat and lifted it as much as 10 feet (3 m) at a distance of 100 feet (30 m). The surcharge had to be removed to another site so that sliding was not reactivated. The final fill has been stable.

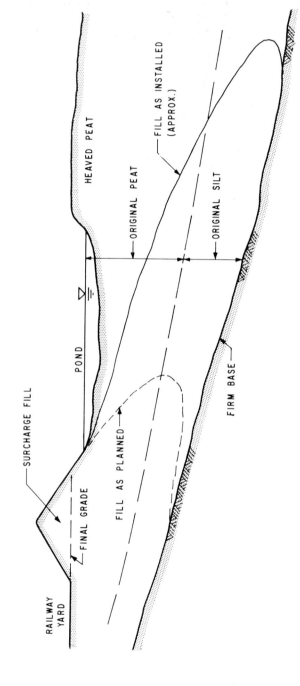

FIGURE 6.25 Section through deposit of peat on silt on sloping base, showing displacement fill as planned and as built

FIGURE 6.26   Displacement fill area at Pontiac with pond formed by consolidation and displacement of peat. Uplifted area is at right

In this case the excessive displacement was caused not only by the slope of the underlying firm surface but by the presence of the soft silt layer. Differences in stress–strain characteristics of peat and soft silt or clay result in the cohesive soils failing before the full strength of the peat is mobilized. Hence, the effective strength of the peat is actually decreased by the presence of weak cohesive soils and a careful engineering analysis of stability is required. More experience with displacement fills on sloping bases is needed before fill quantities can be predicted, even approximately.

The displacement methods reviewed below are also discussed in Section 6.5.

A   Displacement by Gravity

Gravity displacement is the basic method. It is best done with a rolling surcharge at the front end of the fill, advanced with a narrow end, as shown in Figure 6.27. The technique is most successful where the peat is 10–20 feet (3–6 m) deep with low strength.

The effectiveness of gravity displacement can be increased by various techniques. The simplest of these (Casagrande 1966) uses the method shown in Figure 6.28. The surface crust of muskeg is broken up over a width of 5–10 feet (1.5–3 m) along the centre line, preferably by blasting. A narrow fill is then end-dumped along this line followed by the fill over its full width, thus ensuring that a narrow core of fill penetrates deeply into the muskeg. This method can also be used at the start of a fill operation that uses displacement by blasting.

Other techniques to improve displacement are intended either to increase the weight of the fill or decrease the strength of the peat. The weight of a granular fill can be increased in place by jetting water into it until it is nearly saturated. This procedure has been used in combination with a general surcharge to facilitate displacement. The surcharge is later bulldozed to the side.

Methods to decrease the strength of peat are disturbance with water jetting or blasting. Both of these methods help the peat to move out from under the fill more rapidly and completely.

B   Displacement by Jetting

This method is used to break up and soften the peat preferably both before and after the placement of fill. Although effective, it requires suitable pumps, piping and jets, and continuous careful supervision. Sufficient water is usually available.

C   Displacement by Blasting

This method uses explosives to weaken and displace the peat and is much better adapted than jetting to regular construction procedures. Two methods are in use and are described in Section 6.5 and elsewhere (Canadian Industries Limited 1964).

The underfill method is used to muskeg depths of 30 feet (9 m) and more. It may be carried out by putting explosive charges down from the toe of the fill to the required position and then pushing the fill forward before detonation (Casagrande 1966). If further incentive is needed to induce the peat to flow to the sides of the fill, side ditches may be blasted or dug to relieve the lateral pressure.

The toe shooting method (Section 6.5(3)) is used up to depths of 20 feet (6 m) in soft peat and deeper when the technique of torpedo blasting is employed.

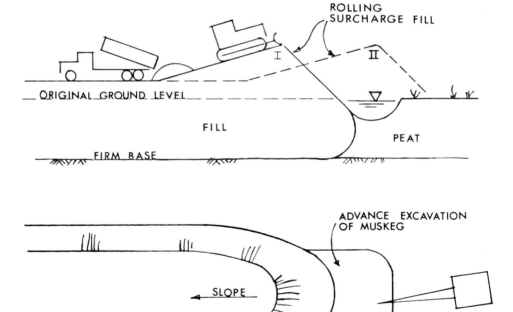

FIGURE 6.27   Gravity displacement method of fill using rolling surcharge and relief excavation at front

(4) *Preconsolidation*

The principle and technique of preconsolidation by surcharging are described in Section 6.5(4). The method is used to accelerate consolidation so that, when the surcharge is removed, the final fill will be stable. The technique requires that the muskeg be sufficiently strong to carry the surcharge load without failure. If this strength is not available, construction can be in incremental stages until full fill height is reached, provided that a factor of safety is present at all times. This stage construction may require many months to complete. The required time is not often available in railway construction.

The preconsolidation method requires a thorough testing and exploration programme, and construction must proceed according to a predetermined schedule confirmed by field measurements. Specialized help is needed in design and in construction control.

(5) *Bridging*

There is a tendency for geotechnical engineers not to admit that a pile-founded structure may, in fact, be the best choice for a muskeg crossing. There are conditions, however, where such a structure is more economical both in first cost and

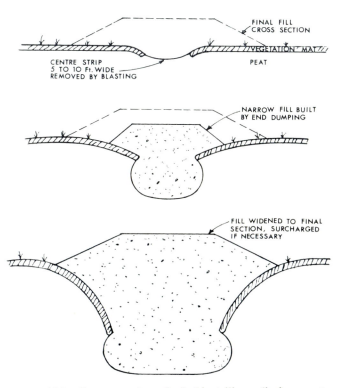

FIGURE 6.28   Centre trench method of installing a displacement fill

maintenance than any type of fill. A pile trestle or bridge should be considered where very deep and weak peat occurs or where the only alternative is a displacement fill on a sloping base. Both these cases are likely to involve excessive quantities of fill. A pile structure may also be indicated where no disturbance of adjacent ground can be allowed, as in crossing close to an existing structure.

Normal procedures are used in exploration, design, and construction for bridges and trestles in muskeg areas. Approach fills are designed as outlined in this Section. It is advisable, however, to place the fill as long as possible in advance of construction of the structure to ensure stability of the abutments or end pile bents.

Replacement of timber pile trestles built to cross muskegs during original construction is now a common railway construction problem. If the railway alignment can be relocated to allow a fill to be placed without disturbing the existing trestle, then the choice of new crossing, whether a fill or a structure, can be based solely on economics. If the alignment cannot be changed and the muskeg is relatively shallow, it may be carefully excavated from around pile bents before a fill is placed. Thorough planning is required before such an operation.

Frequently the alignment cannot be changed and the muskeg is deep. Techniques have not yet been developed for building a displacement fill under these conditions which allow the trestle to carry traffic safely at the same time. Studies of means of doing this, perhaps employing careful jetting, are needed. In the meantime, replacement of the trestle in kind is the only alternative.

(6)   Specialized Procedures
Vertical sand drains and any type of chemical stabilization are not commonly applicable to railway construction on muskeg.

(7)   Construction Costs
The discussion on construction costs in Section 6.6 and the summary of unit costs given in Table 6.3 may be applied to railway construction.

6.22   DRAINAGE

Certain principles govern the effect of muskeg drainage under all conditions.

(a) The submerged weight of peat is only in the order of 5 to 10 pounds per cubic foot (80 to 160 kg per cu m) whereas the saturated weight of the same material when the water table is lowered increases by 62.4 pounds per cubic foot (1 gm per cu cm). This considerable increase in the effective weight of peat caused by drainage greatly increases its rate of consolidation, a factor often overlooked. The effective weight of fill is also increased by drawdown of the water table.

(b) Ditches are effective mainly in carrying away surface water. The water-holding capacity of peat is sufficiently high that ditches have only a local effect in draining the muskeg itself. Evaporation and transpiration will, of course, supplement this drying effect.

(c) Peat undergoes a certain amount of stiffening when drying out, and so actually increases in strength.

All drainage operations should be planned with these factors in mind.

(1) Drainage During Construction

The effects of drainage on various methods of construction can be summarized as follows:

(a) Floating fills – Advance ditching well to each side of the fill area is of benefit to start preconsolidation and strengthen the mat. The effect of the position and depth of the ditches on the fill stability, however, should be checked.

(b) Excavation and replacement method – Drainage may be harmful, promoting undesirable stiffening of the mat and failure of excavation slopes.

(c) Displacement method – Drainage may be harmful, inhibiting displacement of peat.

(d) Preconsolidation method – Water level is controlled to design requirements. The design and installation of surface drainage are fully described in Section 6.7. Airphotos may be used to locate long off-take ditches which are frequently used to reach lower ground. Ditches are excavated mainly by dragline although blasting is sometimes used.

(2) Maintenance Drainage

Increased drainage along the right-of-way is usually thought to be an improvement and sometimes thought to be a cure-all for grade maintenance problems. All drainage, however, should be planned from first principles.

Drawdown of the water table increases the effective weight of the soil and causes consolidation of the fill and any soft underlying material. For this reason, if the fill is resting on firm material, drainage will consolidate the fill material with some benefits. If the fill is underlain by peat as it is in most cases, lowering of the water table will be detrimental by renewing consolidation settlement, and may engender a need for raising of the grade and surfacing of the track.

New ditches should be kept well away from the toe of embankments to avoid shear failures towards the ditch.

6.23 MAINTENANCE OF TRACK

(1) Subsidence of Roadbed

In muskeg territory, maintenance of railway fills is required mainly because of subsidence. The profile of a line develops a sag at each muskeg crossing. This sag alternates with high spots where the grade is on firm ground. Raising of the track may be undertaken in some cases to keep the fills far enough above water to ensure good drainage, and in other cases to smooth out the track profile to allow higher speeds. In either case, the weight of more fill material, if the fill is supported by peat, starts a further cycle of settlement over a period of years. A sound rule is to add as little fill as possible.

The decision whether or not to raise the track should be based on a knowledge of how the fill is supported. If it rests on firm ground, the problem is negligible. If it rests on peat, the alternative of not disturbing that portion but lowering the road-bed on firm ground at each end to smooth out the track profile should be con-sidered. This procedure would maintain the equilibrium of the fill supported on peat. In the same way irregular track may be levelled by lowering high areas rather than by lifting low areas.

Substantial raising of the grade entails the risk of a stability failure. For this reason, grades should never be overlifted with the reduction of future maintenance in mind. If a failure does occur, indicated by a lateral bulge of the peat, the solution is to install a berm, one-third to one-half the height of the fill, to act as a counterweight against further movement. The weight of a berm will, of course, promote further settlement.

(2)   *Bank Widening and Ballast Lifting*
The periodic addition of layers or lifts of fresh ballast is a regular track main-tenance operation. As the total thickness of ballast builds up on top of a fill, the available shoulder width becomes inadequate and further ballast will spill down the slopes. At this stage, bank widening must be done by adding fill material to the slopes, thus increasing the width of the shoulders of fill.

The addition of both ballast and bank materials adds weight to the roadbed and may cause increased subsidence of fills on peat. Since bank widening material is usually hauled by train and dumped down the slopes, it loads the peat along a narrow strip and tends to displace it. The fresh slopes formed are already at their angle of repose, so any yielding of the peat causes the fill to run down and decreases the width of the fill shoulder at track level. An improved practice is to push the dumped material as far down and out as possible with a spreader or bulldozer. In this way the added load is applied over a wider base on the peat and the fill has a better chance at performing its intended function.

(3)   *Running Rail*
The phenomenon known as "running rail" occurs frequently on fills over muskeg areas. Aided by the springiness of the peat, the condition seems to be caused by a general flexing of the track, embankment, and foundation under the load of an advancing train. An actual wave of material is formed, which the locomotive is continually trying to climb. This phenomenon has not been studied in full detail, but one of its characteristics is that rails are actually pushed over the roadbed in the direction of travel. Ties anchored to rails are also carried forward. Figure 6.29

FIGURE 6.29   A case of "running rail" which has moved to the left in the direction of travel of the last train, carrying anchored ties with it

shows a typical location after passage of a train from right to left. Several derailments each year are attributed to running rail. No particular correlation is found with the size of the embankment or the depth of the peat. Movement of the rail relative to the ties can be resisted by the installation of rail anchors; movement of ties relative to the ballast can be resisted by the installation of crushed rock ballast. Heavier rail and increased ballast thickness make the track structure more resistant to flexing. All these measures are helpful in reducing the problem.

REFERENCES

ANDERSON, K. O. and R. C. HAAS. 1962. Construction over muskeg on the Red Deer Bypass. Proc. Eighth Muskeg Res. Conf., NRC, ACSSM Tech. Memo. 74, pp. 31–41.

BRAWNER, C. O. 1958. The muskeg problem in BC highway construction. Proc. Fourth Muskeg Res. Conf., NRC, ACSSM Tech. Memo. 54, pp. 45–53.

———— 1959. Preconsolidation in highway construction over muskeg. Roads and Eng. Const., Vol. 97, No. 9, pp. 99–104.

———— 1960a. Aerial photography in highway studies. The BC Professional Engineer, Vol. 11, No. 12, pp. 11–15.

———— 1960b. The practical application of preconsolidation in highway construction over muskeg. Proc. Sixth Muskeg Res. Conf. NRC, ACSSM Tech. Memo. 67, pp. 13–28.

———— 1963. Flexible pavement evaluation using the Benkelman beam. The BC Professional Engineer, Vol. 12, No. 10, pp. 9–14.

BRAWNER, C. O. and R. O. DARBY. 1962. Culvert selection in muskeg areas. Proc. Eighth Muskeg Res. Conf., NRC, ACSSM Tech. Memo. 74, pp. 84–99.

BROCHU, P. A. and J. J. PARÉ. 1964. Construction de routes sur tourbières dans la Province de Québec. Proc. Ninth Muskeg Res. Conf., NRC, ACSSM Tech. Memo. 81, pp. 74–108.

CANADIAN INDUSTRIES LIMITED. 1964. Blasters' Handbook, fifth ed., pp. 330–335.

CASAGRANDE, L. 1966. Construction of embankments across peaty soils. J. Boston Soc. Civil Engrs., Vol. 53, No. 3, pp. 272–317.

CRONKHITE, R. H. 1967. Highway construction through muskeg areas in northern Alberta. Proc. Twelfth Muskeg Res. Conf., NRC, ACGR Tech. Memo. 90, pp. 93–106.

CUSHING, J. W. and O. L. STOKSTAD. 1934. Methods and costs of peat displacement in highway construction. Proc. Highway Research Board, Vol. 14, Pt. 1, Washington, DC, pp. 325–339.

———— 1935. Methods and cost of filling for highways over swamps. Eng. News-Record, Vol. 114, No. 4, pp. 126–129.

DEWAR, S. 1962. The oldest roads in Britain. The Countryman, Vol. 59, No. 3, pp. 547–555.

ENGINEERING NEWS-RECORD. 1945. Dredges build a road. Vol. 135, No. 4, pp. 114–117.

FLAATE, K. and N. RYGG. 1964. Sawdust as embankment fill on peat bogs. Proc. Ninth Muskeg Res. Conf., NRC, ACSSM Tech. Memo. 81, pp. 136–149.

GODARD, H. P. 1955. The corrosion behaviour of aluminum. Corrosion, Vol. 11, pp. 54–64.

GOODMAN, L. J. and C. N. LEE. 1962. Laboratory and field data on engineering characteristics of some peat soils. Proc. Eighth Muskeg Res. Conf., NRC, ACSSM Tech. Memo. 74, pp. 107–129.

HILLIS, S. F. and C. O. BRAWNER. 1961. The compressibility of peat with reference to major highway construction. Proc. Seventh Muskeg Res. Conf., NRC, ACSSM Tech. Memo. 71, pp. 204–226.

JEFFRIES, J. M. 1936. Building roads through unstable foundations. Civil Eng., Vol. 6, No. 5, pp. 317–320.

LEA, N. D. and C. O. BRAWNER. 1963. Highway design and construction over peat deposits in the lower mainland of British Columbia. Highway Res. Board, Res. Rec., No. 7, Washington, DC, pp. 1–33.

MACFARLANE, I. C. 1956. Techniques of road construction over peat. Roads and Eng. Const., Vol. 94, No. 7, pp. 92–126.

MICKLEBOROUGH, B. W. 1961. Embankment construction in muskeg at Prince Albert. Proc. Seventh Muskeg Res. Conf., NRC, ACSSM Tech. Memo. 71, pp. 164–184.

MONAGHAN, B. M. 1963. Muskeg and the Quebec North Shore and Labrador Railway. Eng. J., Vol. 46, No. 3, pp. 35–40.

MORAN, PROCTOR, MUESER AND RUTLEDGE. 1958. Study of deep soil stabilization by vertical sand drains. U.S. Dept. of the Navy, Bureau of Yards and Docks, Rept. No. y88812, Washington, DC, 429 pp.

PARSONS, A. W. 1939. Accelerated settlement of embankments by blasting. Public Roads, Vol. 20, No. 10, pp. 197–202.

PORTER, O. J. 1939. Stabilization of fill foundations by drainage treatments. Western Const. News, Vol. 14, No. 3, pp. 86–89.

ROMANOFF, M. 1957. Underground corrosion. U.S. Dept. of Commerce, Nat. Bur. Stands., Circ. 579.

TESSIER, G. 1966. Deux exemples – types de construction de routes sur muskegs au Québec. Proc. Eleventh Muskeg Res. Conf., NRC, ACGR Tech. Memo. 87, pp. 92–141.

7 Special Construction Problems

Chapter Co-ordinator *
IVAN C. MACFARLANE

7.1 DAMS AND DYKES

Earth dams and dykes differ from embankments in general in that they are designed to impound water. They must be reasonably watertight, therefore, and for stability the seepage pressures in the body of the fill as well as in the foundations must be controlled, keeping in mind the range in levels over which the impounded water can fluctuate. The immediate and long-term settlement patterns are of importance because of the necessity of adequate freeboard throughout the life of the structure. In addition, excessive differential settlements may crack the dyke embankment and could induce complete failure.

The comparatively high deformation characteristics of peat, even after being highly consolidated, make it hazardous to include it in the body of dam or dyke embankments. Earth dams and dykes can safely be built on peat foundations, however, if proper precautions are taken. The construction of more rigid types of dams or dykes, such as concrete structures, should not be attempted on peat foundations.

While earth dykes are relatively flexible and are capable of accommodating relatively large differential movements, they are much more rigid than peat even after it is highly consolidated. It must be anticipated, therefore, that cracks may develop in such embankments in accordance with their acting as relatively rigid beams resting on a deformable foundation. It is not unusual, of course, for dykes on compressible foundations of inorganic soil to develop cracks. Longitudinal cracks in dyke embankments to heights not exceeding about 20 feet (6.1 m) are not usually objectionable providing the overall stability of the dyke is adequate. Transverse cracks, however, can precipitate failure.

The permeability of peat in its natural state is seldom low enough to keep the seepage loss in the foundation within acceptable limits, or to provide adequate control of seepage pressures. One of the unusual characteristics of peat, however,

*Chapter Consultants: R. M. Hardy – Dams and Dykes; T. A. Harwood; R. A. Hemstock, W. J. Keys – Pipelines; M. Markowsky, G. E. McLure, N. J. McMurtrie   Tower Foundations; N. W. Radforth – Drainage; G. Schlosser – Winter Airstrips; S. Thomson – Frozen Muskeg.

is its extraordinarily large reduction in permeability as consolidation takes place (see Section 4.2(2)). With a reasonable degree of consolidation, the peat foundation can be made sufficiently impermeable for conventional methods of seepage control to be adopted.

A major problem with dykes on peat foundations is the large settlements that must be accommodated as a result of consolidation of the peat and the fact that secondary consolidation (or compression) results in relatively large displacements that are time-dependent. Measurements of pore pressure during and immediately following the application of embankment loads indicate that the pore pressures induced by the load dissipate rapidly. This means that primary consolidation (Wahls 1962) occurs rapidly and therefore an increase in shear strength due to consolidation would also be expected to occur rapidly. Experience in placing fills on peat suggests, however, that a significant increase in the shear strength of the peat occurs after the pore pressures are largely dissipated. This is an anomaly that requires further research attention, but it is of considerable practical importance where stage construction is required to build up sufficient shear strength in the peat to carry the ultimate height of the fill. Failure may be produced in the peat if the rate of loading is based only on the rate of dissipation of pore pressures. Subsequent stages of loading should not be added until the plots of settlement vs. logarithm of time have flattened out virtually to a straight line.

Secondary consolidation (or compression) can be defined as settlement that proceeds with no measurable, or very low, pore pressures existing in the soil. Such settlements generally show a straight line relation between settlement and logarithm of time (see Fig. 4.7). They must be expected to continue, therefore, for the life of the structure but at a gradually decreasing rate. This is true for normal inorganic soils as well as for peat, but some performance records for secondary compression in peat show that the slope of the compression vs. logarithm of time plot increases somewhat with time. Under these circumstances, extrapolation of settlement records to future times would give results that are too low.

The design and construction of dams and dykes require very careful attention to detail if for no other reason than that the release of water following a failure can be catastrophic or result in extraordinarily severe damage. Attention here is confined to the special problems that arise when the foundation material is peat. Within the context of the mechanical properties inherent to peat as noted above, the following factors need to be given special attention in the design and construction of dams and dykes which are intended to be supported on the consolidated peat rather than to displace it.

(1) *Stability of Foundation*

Mathematical analysis of the stability of peat foundations requires estimates of the shear strength characteristics of the peat as discussed in Section 4.4(1). In particular, knowledge of the *in situ* shear strength, angle of internal friction, and K_o values are required. Because of the great difference in the deformation characteristics of the peat as compared with the dam or dyke fill material, stability analyses should assume that the shear resistance of the embankment contributes nothing to the overall stability. The peat itself must be capable of resisting the total shear forces. The effect of reducing K_o value with increasing effective consolidating pressure requires special attention.

These considerations will dictate whether stage construction is necessary, the allowable maximum height of the various stages, and the minimum width of base of the embankment. They will usually dictate the use of berms to widen the base apart from what might be required for the stability of the fill itself. The use of berms is preferable to flattening the slopes of the fill to the overall average slope from the toe to the shoulder of the dyke, because it is desirable to keep the total weight of the applied fill to a minimum. The stability at the end of construction, and with water ponded to full supply level, also requires special attention in view of the seepage pressures that may be induced from the ponded water.

(2)   Settlements

The question of peat consolidation and settlement is discussed in detail in Section 4.4(5). In the present Section, the theoretical background is presented for the special case of design of dykes on peat foundations. This approach differs slightly from that of Chapter 4, but the discussion that follows represents an application of the present knowledge of peat consolidation characteristics which has resulted in the successful construction of dykes on muskeg.

Settlement in soil types in general is a combination of primary consolidation and secondary compression (Wahls 1962). For peat in contrast to most – but by no means all – inorganic soils, the magnitude of the secondary compression may exceed the primary consolidation within periods of time less than the life of the structure. Moreover, in plots of settlement vs. logarithm of time for either laboratory or field data, the secondary compression may almost completely obliterate the primary consolidation.

In these circumstances, it is impossible at present to separate accurately primary consolidation and secondary compression as is done in conventional design practice with most inorganic soils. Neither is it possible to make meaningful analyses of the rate of consolidation based on primary consolidation theory. The usual practice in attempting to estimate settlements of embankments on peat, however, has been to apply the theory of primary consolidation using the relation

$$S_0 = H_0 \frac{C_c}{1 + e} \log_{10} \frac{p_0 + \Delta p}{p_0}, \tag{1}$$

where S_0 is the settlement, H_0 is the thickness of the compressible layer, e is the void ratio of the peat, p_0 is the initial load on the peat, Δp is the consolidating load increment, and C_c is the compression index of the peat.

The value of C_c is a soil characteristic that can be determined from standard laboratory consolidation tests on the peat, or from field settlement observations.

It is difficult to secure accurate values for C_c from laboratory tests on peat, but it is known from both laboratory and field test data that it is usually considerably higher than for normal types of inorganic soil. Depending upon the accuracy with which C_c values can be estimated, equation (1) will give realistic estimates of settlement, providing the C_c value used has been derived from settlements taken at times for which the major portion of the settlement under the particular load increment has been completed. A fundamental error in the use of equation (1), however, is the fact that C_c is representative of conditions at a finite time, or a series of finite times, while, in fact, if secondary compression is predominant in the soil, the magnitude of the settlement is predominantly time-dependent. In dealing with peat, estimates of settlement which ignore the time-dependent nature of the movements can lead to serious misconceptions of the ultimate performance of a structure.

An alternative to the use of equation (1) is to base the settlement analysis on the assumption that the settlement has two components: an instantaneous fraction which occurs immediately on application of the load, followed by a second component which proceeds in accordance with a straightline relation between settlement and logarithm of time. The instantaneous fraction can be estimated by using equation (1) with an appropriate adjustment in the C_c value, or alternatively a direct extrapolation can be made from pressure-void ratio curves from laboratory consolidation tests. For the second component the equation for secondary compression can be used, namely

$$S_s = H \frac{C_s}{1 + e} \log_{10} \frac{t}{t_0}, \tag{2}$$

where S_s is the settlement within the time interval t_o to t, H is the thickness of the compressible layer, and C_s is the coefficient of secondary compression expressed in terms of change in void ratio over one cycle of the logarithmic scale on a plot of void ratio vs. logarithm of time.

The value of C_s can be determined from laboratory consolidation tests or from field settlement data in the same way as values for C_c, but its determination is subject to the same difficulties noted for the evaluation of C_c.

Field settlement data are usually plotted in terms of settlement vs. logarithm of time. In this case, the settlement over one cycle of the logarithmic scale becomes the term $HC_s/(1 + e)$ in equation (2).

The physical mechanism causing secondary compression is not clearly understood, but there is considerable evidence suggesting that C_s is not a constant soil parameter. A major factor affecting its value appears to be the magnitude of shear stress in the soil. This can readily be shown in laboratory tests in which samples are consolidated in triaxial compression tests under different intensities of shear stress. Under conditions of zero shear stress, C_s is zero, but it increases with increasing shear stress in the soil (Hunter 1967). Results such as these suggest that the standard laboratory consolidation test is not the best method for evaluating C_s, because the shear stresses in the soil specimen are unknown and they may vary with the consolidating pressure.

More sophisticated mathematical analyses have been made (Wahls 1962) by combining equations (1) and (2). The accuracy of the values that are assigned to C_c and C_s, however, is still the major factor governing the precision of the settlement predictions. The best that can be expected, in the light of present knowledge, is that settlement analyses based on values of C_c and C_s from laboratory test results will give only a rather crude estimate of the magnitudes of the settlements to be expected. For more precise estimates of the ultimate settlements, the results of settlement observations made during and immediately following construction on the actual job must be used.

Settlement concepts are also essential in assessing the hazard of transverse cracking in an embankment due to differential movements. Critical sections are to be expected above areas of abrupt changes in foundation soil profiles. These may exist because of large variations in the thickness of peat over short distances along the dyke alignment, or because of an abrupt change in soil type, such as that from peat to an inorganic soil.

With relatively thick peat deposits, attention may need to be given to the horizontal components of the consolidating stresses. These stresses have a tendency to induce a spreading failure in the embankment. The use of berms in the dyke cross-section produces a more favourable stress distribution from the point of view of horizontal consolidating forces.

(3) *Seepage*

The principles of seepage control for dams and dykes are for the most part directly applicable to those on peat foundations. Because of the substantial decrease in the permeability of peat as it is consolidated, the problem of the control of seepage through the foundation soil is less severe than for many more normal soil types.

Standard facilities for seepage control such as upstream impervious blankets, foundation cut-offs, internal drains, and downstream seepage relief wells are all applicable to peat foundation conditions. A major special problem arises where vertical seepage cut-offs, or vertical seepage pressure relief wells are installed in the peat. These facilities are usually much more rigid and considerably less compressible than the surrounding peat. Consolidation of the peat subsequent to the installation of such facilities results in excessive loads being transmitted to them. The facilities themselves may rupture and fail. Alternatively, in the case of vertical cut-offs below the dyke, they may act as a rigid "plug" below the embankment which may result in the embankment being broken. Longitudinal cracks may develop in the dyke on either side of the cut-off. These may or may not be hazardous, but they do result in a reduction in the overall stability of the structure.

The use of vertical cut-offs through peat is a carryover from their use in relatively pervious inorganic soils. They are often used in peat because the substantial decrease in permeability due to consolidation is not recognized. It most cases, therefore, they will be unnecessary. Where they are used, they should not be placed below the centre of the dam or dyke. Rather, their position should be either well upstream or downstream of the centre line, so that, if longitudinal cracks do develop because of differential settlement, they will not affect the main body of the dyke.

(4)   Embankment Materials

No special problems exist with the selection of materials for the embankment where a dam or dyke is to be built on a peat foundation. There is, however, one consideration. This arises where the differential settlements are expected to be such that transverse cracks may be produced in the embankment. Homogeneous fills of highly impervious clays or silty clays have been successfully used for dykes on muskeg, but these have the poorest self-sealing characteristics. A zoned embankment with properly designed transition sections will provide the maximum self-sealing properties. A zoned embankment includes an impervious core section surrounded on each side by more pervious sections, preferably with granular material being used for the outside zones. The effectiveness of such a cross-section in inducing self-sealing of cracks which may develop in the impervious core depends on the efficiency of the transition zones that should be provided between the adjacent zones of different permeabilities. Records are available (Hardy 1967) showing that dams with zoned cross-sections and properly designed transition sections are self-sealing with transverse cracks that have opened up as much as 3 or 4 inches (7.6 or 10.2 cm). The use of thin impervious zones to control seepage through the dyke should be avoided with dykes on peat foundations.

With homogeneous cross-sections, water should not be permitted to rise against the upstream face if transverse cracks are showing on it, particularly if portions of the fill are frozen. The material in the upstream face should be re-worked to seal off the cracks before the water level is raised against the dyke.

Properly designed internal drains, particularly if they are vertically continuous in the embankment, also provide good self-sealing characteristics. They are unlikely to be as efficient, however, as a properly designed zoned cross-section.

Under some circumstances, horizontal cracks may develop in an embankment. The circumstances most likely to produce such cracks exist where relatively deep frost penetration develops below the surface of the dyke, while at the same time settlements are developing in the peat foundation. Such horizontal cracks could precipitate failure if water is brought up against the crack before the crust has thawed out. The cracks will usually be self-sealing after the frozen crust has thawed.

Longitudinal cracks, either along the upstream or downstream faces, usually are not objectionable. They can be expected to be self-sealing provided they are caused by consolidation in the underlying peat rather than by a partial shear failure in the peat. If the vertical displacement at the crack exceeds a few inches, it is likely that a partial shear failure has occurred. This condition would dictate widening the base of the fill in such areas by using a berm.

(5) Construction Details

For dam or dyke construction on peat, the trees and shrubs should be cleared from the area of the base. Contrary to common practice with certain other types of embankments on peat, the trees and shrubs should not be incorporated into the base of the fill as "corduroying," because a seepage failure path may develop through loose soil surrounding the logs or brush. The larger roots should be dug out, but the surface layer of live peat need not be stripped off.

The initial lift of the fill is usually placed by end-dumping to secure a mat on which the earth hauling and compaction equipment can operate. This initial mat should be 2–4 feet (0.61–1.22 m) thick. It is impossible to secure a high degree of compaction in this initial lift. One of the several reasons for the use of berms to widen the base of dams or dykes is to compensate for the poorer quality of fill that will probably form this initial lift. Compaction should be started, however, as soon as possible. For the construction of any sort of dam or dyke it is essential that compaction procedures be used, but the emphasis should be on the securing of a uniform quality of fill rather than on high densities. In the lower portions of the embankment, over-compaction in attempts to secure high densities can result in objectionable disturbance to the underlying peat. Compaction on any lift should be stopped temporarily if the equipment begins to produce appreciable "weaving" due to disturbance of the underlying peat.

Drainage of the muskeg area, for the purpose of lowering the water table at the base of the embankment, can be beneficial in that some consolidation of the peat will be induced prior to placement of the fill, and a thinner initial mat may be required for the fill. It is preferable that the drainage ditches be directed away from the dyke area at right angles to the dyke axis. If drainage ditches must be located on lines more or less parallel to the toes of the dyke, they should be kept as far outside the toes as practicable. Drainage ditches along the toes of the dyke seldom have practical advantages, and they are detrimental to the stability of the peat.

Construction of dams and dykes in winter has several advantages if the operations are properly conducted. While it may be necessary to remove or to compact the snow cover before the peat will freeze, once it is frozen the operations of clearing and stripping can be done more economically than in the summer. Drainage ditches, or seepage cut-off trenches in the peat below the dyke, can also be dug more readily in frozen peat. As a rule a backhoe can cut the frozen peat, and the trench can be dug with vertical sides. This eliminates the difficulties of access and excavation below the water table inherent in summer operations.

Placement of the initial lift of the embankment and backfilling for cut-off trenches can also be done more readily when the peat is frozen. Satisfactory compaction can be secured in these operations if the fill material is excavated in the borrow area in an unfrozen condition and then placed and compacted before it freezes. This can usually be accomplished without too much difficulty at air temperatures down to about $10°$ F $(-12°$ C), and there are records of it being done at even lower air temperatures (Hardy 1967). With efficient organization of the work and with favourable air temperatures, higher quality of backfilling and of the initial lift in the fill may be achieved than is practically possible under summer conditions.

With embankments constructed on frozen peat, the question arises as to the stability conditions when it thaws. If the peat should thaw instantaneously, or over a very short period, failure of the peat foundation would be expected. Experience has shown, however, that peat thaws very slowly, and that it may, in fact, take several seasons before it completely thaws below the embankment. Under conditions of slow thawing, local consolidation and increase in strength appear to occur in the soil, so that overall instability does not develop. Shortly after thawing is complete, the stability and consolidation conditions are not greatly different than if no freezing had occurred. The settlement pattern during the thawing period will be distorted, however, and the usefulness of field settlement data in extrapolating to future settlements, or in checking laboratory test values for C_c and C_s, may be lost.

Where stage construction procedures have been necessary, experience indicates that there is less chance of failure in the peat when the first stage is built while it is frozen. This suggests that stage construction might not be necessary if the fill is placed on frozen peat. There are insufficient performance data available, however – and a lack of methods for rationally analysing the conditions, as well – to warrant abandoning stage construction procedures where they are dictated by the summer season construction conditions.

(6)   Displacement of Peat

The practice of displacing the peat by the weight of the fill or by blasting to secure a firm foundation for an embankment has been widely used. Consideration of the techniques used with these procedures is beyond the scope of this discussion. They are dealt with in some detail, however, by Casagrande (1966) and in Section 6.5(3) of this Handbook.

7.2   AIRSTRIPS

Construction of airstrips in areas of organic terrain falls naturally into two main classes which will be considered separately: (1) conventional airstrip or airport runway construction as a part of an existing airport or the development of a new airport; and (2) winter airstrips on muskeg which are built and used only on a seasonal basis.

(1)   *Conventional Airstrips*

For the most part, peat areas are avoided in the construction of airstrips and air-port runways, but sometimes it is unavoidable, particularly in the case of a runway extension. Muskeg areas have the advantage, however, of being flat and normally free from rock outcrops, although they do require special treatment to make them stable. By far the most common means of eliminating trouble is to dispose of the peat by excavation, hauling it from the site, and backfilling with select granular material. In this regard many of the principles outlined in Section 6.5(2) for highways can be followed in the construction of airstrips and airport runways. The extent of excavation is generally greater than for a highway, however, and the choice of a method of excavation is critical. This will depend upon a number of factors, including economic considerations, location, amount of peat to be removed, and equipment available. Prior to excavation operations, drainage of the site is advisable to avoid difficulty with excavation equipment.

In deep muskeg areas, draglines are normally used for excavation (Highways and Bridges 1942; Roads and Streets 1944), with a clamshell bucket often favoured over the dragline bucket (Engineering News-Record 1942; Stanwood 1958). In shallow deposits and where the peat is dry enough, a bulldozer can be used to assist in the removal of the peat. It is piled in windrows for loading by clam-shell bucket for haul to the disposal area. Bulldozers were used in a most unusual way for removing peat in the construction of the Digby Island airport near Prince Rupert, BC (Stanwood 1958). In one location the muskeg overlaid rock which sloped rather sharply away from the runway area. Bulldozers, pitted against a 12-foot (3.66-m) face of peat, pushed downhill and were able to move the mass completely off the runway area. Before the peat mass could build up and bottle-neck, it turned into a fluid and flowed like lava.

A second and rather less common method of construction is to stabilize the peat by preconsolidation, a technique that has been used for a number of years (Munoz 1948; Kapp *et al.* 1966). The principles outlined in Section 6.5(4) for preconsolidation in highway construction are equally applicable to airport runway construction. A very careful and thorough field and laboratory investigation must necessarily precede this construction method.

(2)   *Winter Airstrips on Muskeg*

The construction and use of airstrips on the surface of muskeg are largely restricted to the winter. Frigid conditions with frequent snowfalls provide both the firm base and the material for surfacing the runway. Winter airstrips can be built inexpensively and are easily maintained. In addition, the costly destruction of merchantable timber is avoided.

Since the use of aircraft in remote or bush operations plays such a vital part in the economy and efficiency of these operations, it is desirable to have the landing strips as near as possible to the operations site. Mineral terrain suitable for a landing strip and sizeable lakes are not always convenient to the site so that a strip on muskeg is often the only alternative.

There are two categories of winter airstrips on muskeg: (i) the short strip for light aircraft used in servicing a mobile operation, and (ii) the long strip used for medium-range aircraft in airlifting cargo and personnel to a winter-long operation.

A   Site Selection

The selection of a site for an airstrip on muskeg can be made either from a study of large-scale maps and air photographs or directly by an aerial reconnaissance over the area surrounding the operations site. The map–airphotograph method is more practical. A number of possible sites can be selected, located, measured for detail, and thoroughly examined – all inexpensively. Often subtle details recognized when flying over the site at 70 miles (113 km) per hour can be identified in this way.

In the investigation, the photo interpreter first seeks out long, open muskeg areas which are relatively free of tree cover or contain a cover of low bushes and shrubs. The muskeg types in this cover class are FI, EFI, and DFI (see Chapter 2). These types are almost always associated with the drainage pattern of large muskeg masses and are distinguishable on the airphotograph by the light grey to nearly white tone and uniformly flat texture on high-altitude photographs.

The topographic features associated with these cover classes in this particular regime are hummocks or tussocks, peat mounds, and peat ridges. The latter feature does not occur too frequently in these situations. All of these are readily removed without destroying the even plane of the muskeg surface. The site selected must be free of large obstacles such as hills and dense stands of high trees on the approach to the strip at each end.

B   Construction

Where the site has been selected by a photo study or by aerial reconnaissance it must be thoroughly investigated on the ground. An appraisal is made of the ground conditions along the entire length of the line for frequency of occurrence of mounds and ridges, the presence of stream channels, ice-filled voids in the muskeg surface, and undulations. Stream channels, voids where ice is not frozen to the bottom and the depth from the muskeg surface to the top of the ice is more than 3 inches (7.62 cm), and undulations of 3 feet (0.91 m) in height over a distance of 500 feet (152.5 m) must be avoided.

With a bulldozer, the centre line of the proposed runway is cut and cleared, bare, one blade width, to the surface of the muskeg. If conditions on the centre line appear to be safe for use as the site of a runway, then two additional lines are cut in the same manner, 50 feet (15.25 m) on either side of the centre line. These are also thoroughly inspected. The conditions on the three lines may be such that a shift in alignment or extension beyond either end, rather than relocation of the strip, may be required.

Muskeg strips can be cleared to a minimum width of 150 feet (45.8 m) – 75 feet (22.9 m) on either side of the centre line of the runway – with the runway 75 feet (22.9 m) wide and 2000 feet (0.61 km) long for category 1 airstrips. For single engine aircraft with STOL* abilities, these are minimum requirements. For category 2 airstrips which must service medium-range aircraft, the minimum width for cleared strips is 200 feet (61.0 m) – 100 feet (30.5 m) on each side of the centre line of the runway – with the runway 100 feet (30.5 m) wide and 5000 feet (1.53 km) long.

In clearing the site, especially the runway, it is advisable to have the bulldozers shear the surface of the muskeg with their blades to pare off any trees and shrubs at the surface. This also helps shave off the tops of the tussocks, mounds, and ridges.

Once the snow and vegetation have been removed to the sides of the strip, the surface can be levelled. Any mounds or ridges must be removed by scarifying or ripping and paring down to the level of the runway with a rear-mounted hydraulic ripper on a bulldozer. The tops of undulations can also be pared off by this method.

After the levelling has been completed, the runway surface is made smooth by blading snow onto the runway and spreading it over the surface with a heavy metal drag. The smoothing may have to be done in stages to permit the processed snow to set up and age-harden. In low humidity areas the addition of moisture by spray from a tank truck will aid the age-hardening process and ultimately hasten smoothing of the runway surface.

The prime requisite for airstrips in muskeg is cold weather before, during, and after the preparation of the airstrip. Once prepared, a strip can be made ready for service much earlier in successive winter seasons with only a minimum of work – essentially smoothing the runway surface by compacting the snow rather than by removing it and then allowing it to age-harden.

Winter airstrips on muskeg are convenient, inexpensive, and as serviceable as strips on mineral soil terrain.

7.3   BUILDING FOUNDATIONS

The construction of buildings on peat presents numerous engineering problems – chiefly settlement – which are not too dissimilar to those encountered for roads and other embankments on peat. It also presents an economic problem, since construction costs in peat will usually be greater than those in firm inorganic soils. Under certain circumstances, however, this problem may be more than balanced by the cheaper cost of peatlands, which may remain unused long after all adjoining mineral terrain has been utilized.

Difficulties in building on peat can arise if inadequate precautions are taken with design and construction; some experiences have been both unfortunate and costly

*Short take-off and landing.

(see Fig. 1.25). In general, peat areas have been avoided, but with rapidly increasing land costs in urban areas, the reclamation and extensive utilization of formerly marginal land is becoming an economic necessity (FENCO 1960). The New York Port Authority, for example, has reclaimed marshlands (containing layers of organic silt and of peat) for such diverse projects as runways, buildings, roadways, bridges, open paved areas, wharfs, and water storage tanks (Kapp *et al.* 1966). In some areas of Northern Canada, the rapidly accelerating pace of development has required an increasing amount of construction in regions of both continuous and confined muskeg.

For any particular muskeg site, a careful engineering study and analysis is a prerequisite to determining the cost in advance of construction. The drainage characteristics of the site must be assessed. The exact depth and physical characteristics of the peat, as well as the depth and physical characteristics of any soft inorganic soil underlying the peat (which may create more of a problem than the peat itself), must be determined. This will require more than a casual field and laboratory investigation.

When all pertinent site information is at hand, the engineering analysis will consider a number of possible designs. The selection of a particular treatment for the peat depends on the structure to be built, as well as on the soil conditions. Some methods of treating peat for building construction include the following:

(1)   *Complete Removal of the Peat and Backfilling with Granular Fill*
This method obviously is feasible mainly when the peat is shallow, and when it is underlain by a firm mineral subsoil or rock. The peat is replaced by a properly compacted granular fill and building foundations are constructed in the conventional way. Martin (1965) reports that the technique for construction of houses on shallow muskeg (up to 8 feet (2.44 m) deep) in Prince Rupert, BC is to excavate to solid bottom (which is always rock at that location), and then to use ringwall or pier foundations.

(2)   *End-Bearing Piles*
For deep peat deposits and for heavy structures, this technique has been commonly used as a successful method of overcoming the problem (Goodman and Lee 1962). The peat, in effect, is bypassed completely and standard pile design formulae can be utilized. Even so, difficulties can arise. A building in Northern Ontario – a new car showroom and garage – where this technique, with modifications, was used, was built over a deep peat deposit. Long end-bearing piles to the underlying rock were used under the exterior walls and interior columns. The floor was designed, however, so that floor loads were taken by a granular fill spread over the surface of the peat between the wall and column footings, with little thought being given, apparently, to the high compressibility of peat. Although the building itself was stable, within a year of construction the centre of the garage floor had settled at least a foot, with rather unfortunate results.

(3)   *Friction Piles*

This technique is advisable only for light buildings in very deep muskeg (see Fig. 1.12). The load on the pile is transferred to the soil through skin friction. The length, size, and number of piles required can be determined only by driving test piles and making the appropriate calculations.

Martin (1965) reports that in Prince Rupert, BC, for buildings of less than 125 pounds per square foot (610 kg per sq m) floor load (private homes, small warehouses, etc.), the practice is to: (a) excavate below the water table; (b) drive pin piling 2 feet (0.61 m) on centre, battered (driven with an air jack, hydraulic jack, or air hammer); (c) reinforce and cap with concrete; (d) use ringwall construction and develop for piers. Over a 10-year period for 48-inch (1.22-m) ring-wall and 72 by 72 inch (1.83 by 1.83 m) pier footings, Martin observed that settlement was less than one-eighth of an inch (32 mm), since the pin piles had developed some friction value. It should be pointed out, however, that a common peat type in the Prince Rupert area is Radforth category 17 (see Chapter 2), which includes large woody elements up to tree trunk size. Consequently, it provides more stability than many other peat types.

When piles are used for private houses or other small buildings, they are usually made of wood, either treated or untreated. It is imperative, therefore, that the water level be maintained; if it fluctuates or is lowered, the piles will deteriorate.

(4)   *Raft or Slab Foundations*

In this method the building is supported on a compacted granular fill spread over the surface of the muskeg. Although this technique has sometimes been used, it is not recommended since differential settlement will almost certainly occur (Gustafson 1965) with the ensuing difficulties that this entails.

(5)   *Site Preloading*

This method consists of applying a surcharge to the site to develop settlement prior to construction of the proposed structure. Because total peat compression consists of a large element of secondary compression, this, as well as primary consolidation, must be taken into account. The New York Port Authority has utilized this technique very successfully in the construction of warehouses and storage tanks in reclaimed marshland areas (Kapp *et al.* 1966). Goodman and Lee (1962) report the successful use of preconstruction surcharge for a truck garage in Syracuse, NY.

In the design of an appropriate surcharge, this technique can be applied as outlined for preloading in Section 4.4(5)D or for preconsolidation for highways in Section 6.5(4). Depending upon local soil conditions, such features as lightweight fill (slag, sawdust, wood chips, etc.) and sand drains (where appropriate) may be incorporated. Usually, the latter will be desirable when the peat occurs in combination with an underlying soft inorganic soil area.

7.4 TRANSMISSION LINE STRUCTURE FOUNDATIONS

Although transmission engineers tend to avoid areas of weak soil conditions, they are sometimes forced to employ them by either one or both of two significant economic facts. First, transmission lines are often allocated to land not generally desirable for other uses. Second, a straight line is usually the most economical route. Angle structures are much more expensive than the "tangent" structures used on the straight-line portions of a transmission line. Therefore, many lines are located partly in swamp or muskeg areas. Transmission engineers expend considerable effort in looking for foundation designs that will minimize the higher costs of these applications. The information presented here represents experience gained in the construction of transmission line structure foundations in the provinces of Ontario and Manitoba, although it is applicable to similar terrain conditions in any area.

Although in the past many long-distance transmission lines have crossed areas of swamp, muskeg, and marine clay, the continuing development of remote sites presents special problems due to the greater extent of the muskeg to be traversed. The high cost of transporting suitable conventional materials challenges engineers to seek alternative designs which will minimize the need for these materials. Furthermore, the design of transmission towers is concerned not only with bearing pressures but also with overturning forces which result in uplift reactions and horizontal forces on the foundation.

(1) *Nature of the Structural Loading*
High-voltage transmission lines are usually supported by rigid structures but may also make use of semi-flexible ones. The rigid towers are almost always made of steel, while the semi-flexible structures can be made of either wood or steel, with wood predominating in practice.

Transmission line structures can also be classified into three broad groups by the type of loading or duty to which they are applied. On straight-line portions "tangent" or "suspension" structures are used. These structures are loaded mainly through the action of the wind on the bare conductors, or on ice-covered conductors. Because of this, the normal everyday load is a relatively small percentage of the ultimate design load and is the same on each footing of the foundation.

A second category consists of "medium angle" or "semi-anchor" structures which are used at most of the angles in the line. Because the conductor tension normally amounts to several thousand pounds, the transverse component of this line tension at line angles adds considerably to the structural loading of both the tower and its foundation. At heavy angles and line terminations, a third category of structure known as a "heavy anchor" structure is used. This type is used to support the full line tension of all conductors. At semi-anchor and heavy anchor locations, where the conductor tension forms a major part of the structural loading, the everyday loading is a large percentage of the ultimate design loading and, for this reason, transmission engineers take few chances in the foundation design of such structures.

Because of the obvious difference in cost between these structure types (varying in ratio from 3/1 to 6/1), transmission lines are constructed with suspension towers wherever possible. The unusual or unorthodox approaches to foundation design are limited to this type of tower, but because it comprises 95 per cent of the structures used in a typical line, the savings achieved by special approaches can be multiplied many times over throughout the length of the route.

The structural loading is transmitted to the towers and foundations by the action of wind and ice on the conductors; therefore there are problems in dealing with the stresses in the soils from bearing, uplift, and horizontal shear forces. "Suspension" structures, which are located in a relatively straight line, are therefore loaded almost entirely by the overturning force of the wind either on the bare conductors or on ice-covered conductors. This complicates the foundation design to the extent that the backfill material becomes just as important as the soil under the foundation.

(2)   *Extent of Soil Investigation*

Because of the relatively low cost of a transmission line structure and foundation, it is usually difficult to justify extensive soil investigations at each site. For example, the cost of a complete soil investigation would approach the cost of a normal suspension tower foundation in good soil. Consequently, soils are examined visually during the process of selecting a route. The depth of the poorer soils encountered is determined by the probe method. Tower locations in poor soil areas are investigated by the standard drop hammer test but without laboratory analysis.

It should also be borne in mind that the limitations on uniform settlement on the foundation of the transmission line structure are not severe. Differential settlement, however is a major concern because of the additional stresses which this could introduce into the structure. Differential settlement on rigid transmission line structures is limited to ¾–1 inch (1.9–2.5 cm).

(3)   *Design of Tower Foundations*

A   Ontario Hydro Practice (Markowsky and McMurtrie 1961)
(i) SOIL STRENGTHS AND TREATMENT IN DESIGN   In dealing with weak soil areas in Ontario, the unconventional foundation designs are applied only to the suspension or tangent towers because, unlike the semi-anchor and heavy anchor towers, the normal everyday load on these structures is a relatively small percentage of the ultimate design load. Normally, the standard steel grillage foundation of a suspension tower is used for soils with a bearing capacity of the order of 4000 pounds per square foot (1.95 kg per sq cm). In isolated instances, however, these foundations would be used in soils with bearing strength as low as 2000

pounds per square foot (0.98 kg per sq cm). For the semi-anchor and heavy anchor towers, this value is limited in all instances to 4000 pounds per square foot (1.95 kg per sq cm).

When the actual soil conditions at the construction site fall somewhat below these values, the bearing area under each footing is increased by a reinforced concrete mat. This device is used for soil strength down to 1000 pounds per square foot (0.49 kg per sq cm) in the case of suspension towers and down to 2000 pounds per square foot (0.98 kg per sq cm) in the case of semi-anchor and heavy anchor towers. For soils below the strength of 2000 pounds per square foot (0.98 kg per sq cm), piles are used for the foundations of anchor towers. In the case of suspension towers, where the soil strength is below 1000 pounds per square foot (0.49 kg per sq cm), piling is usually considered as a next step. Since access problems, however, frequently make this economically unfeasible, alternatives are sought. A number of these alternative designs which have been used for suspension tower foundations can be called unconventional.

(ii) SPECIAL TYPES OF FOUNDATIONS   A "piston" foundation is shown in Figure 7.1. The significant feature of this type of foundation is the use of cribbing or shoring for the purpose of containing the soil down to a depth at which the stress in the soil has been reduced to a safe figure. A reinforced concrete mat is also used in order to spread out the stress inside the "piston." Two sections of shoring are used, as shown. The granular material which is placed between the inside and the outside shoring is used to increase the friction between these planes for the purpose of "keying" the footing into the plane of the outside shoring. In this way, all of the backfill used will contribute to the uplift strength of the footing. The gravel or planking under the reinforced concrete mat provides a working surface.

A spread mat foundation consists of the steel grillage tower footing on a reinforced concrete mat, and may look very similar to the mat shown as part of the piston footing in Figure 7.1. This method is used primarily to deal with a bearing problem. It is applied where the soil above the footing has sufficient resistance to provide the uplift strength required. A variation of this method, where a smaller increase in bearing area is required over the standard footing, uses ordinary wood planking instead of reinforced concrete. The value of this latter version is limited by the relatively smaller area over which it can be used.

In some cases, soil investigations reveal the presence of relatively thin layers of stronger soils at certain levels. Under these conditions, it is sometimes possible to adjust the depth of the tower footing to a level at which advantage can be taken of one of these stronger layers. In such cases, the spread mat is used together with the top or outside shoring shown in Figure 7.1.

It is sometimes possible to consolidate a weak soil by placing rock or broken concrete at the bottom of the foundation excavation. This has been used to advantage to build up a sufficiently thick layer of stronger material. The footing is then treated in the same manner as above. In both of the cases outlined, the size of the reinforced concrete mat is selected so that the stresses reaching the weaker layers underneath are within permissible limits.

A raft or mat foundation, which has been used to some extent (especially in Northern Ontario) on twin wood pole structures, seems to be particularly adaptable to muskeg conditions. This foundation is illustrated in Figure 7.2. The bearing stresses are distributed over the surface of the ground by the wood planking which is formed into the shape of a raft by the cross timbers. In muskeg, this type of foundation makes very good use of the top dense fibrous layer which, of course, is the only part of the muskeg worth using for this purpose. As a necessary supplement to this type of foundation, the wood pole structures are guyed to each side and are held by swamp or muskeg anchors of the type shown in Figures 7.3 and 7.4. The screw anchor is employed where the swamp or muskeg is not excessively deep and where the blade of the screw can be turned sufficiently far into firm soil to give the desired holding power. The log anchor, on the other hand, is used where

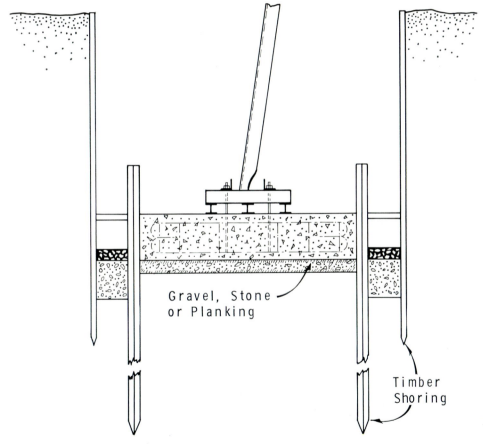

Gravel, Stone
or Planking

Timber
Shoring

FIGURE 7.1   Piston foundation

firm soil is not available at a reasonable depth. In this case, the use of logs and stakes spreads the stress over a large area of undisturbed material.

In the applications described above, the piston-type footing has been used to a limited extent, but quite successfully. The other procedures have been used widely and at considerable savings in cost over the more conventional pile foundations.

A relatively new type of footing for towers is illustrated in Figure 7.5. This footing consists of conventional steel grillage with the addition of bracing to give the necessary lateral support, and earth anchors to give the necessary uplift strength. As shown in the illustration, the peat in this area is relatively shallow. The glacial till which underlies it has sufficient strength to support the tower adequately.

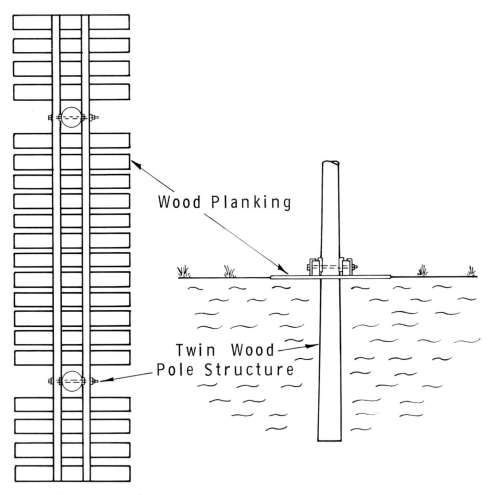

FIGURE 7.2   Raft foundation

The peat which has been removed from the excavation, however, would supply very little support in uplift. As an alternative to the expensive process of importing suitable backfill material, additional bracing and earth anchors, as illustrated, are used.

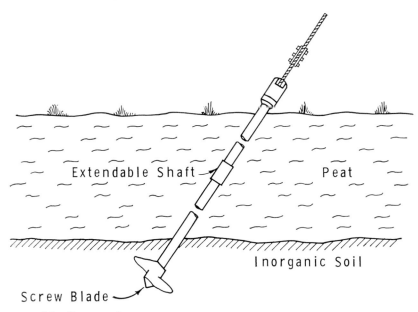

FIGURE 7.3   Screw anchor

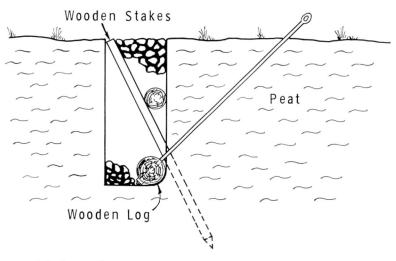

FIGURE 7.4   Log anchor

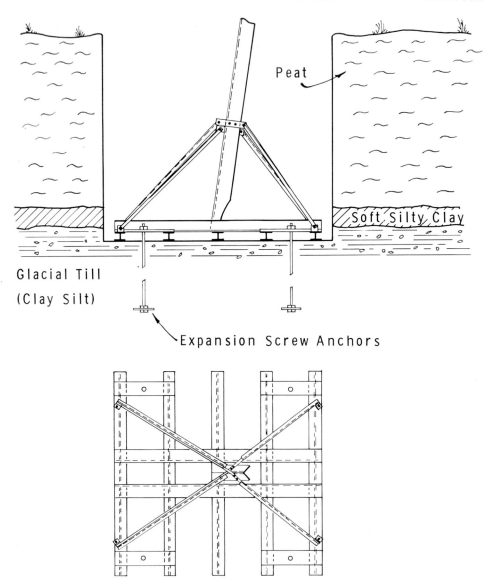

FIGURE 7.5   Steel grillage with anchors

(iii) FROST PROBLEMS   Over wide areas of Northern Ontario, the soil underlying the muskeg, or adjacent to it, varies from pure silt to clay silt. This type of soil is very susceptible to frost heaving and has caused considerable trouble to transmission line structures. Where this soil is overlaid with peat, penetration of the frost is prevented by the insulating muskeg blanket.

When excavating in shallow muskeg, it has been found that, if the silt is mixed with the peat in the backfilling operation, the frost will penetrate the silt mixture into the underlying silt and cause severe heaving. Therefore, precautions are taken to replace the insulating blanket of muskeg.

Where no muskeg cover exists, there are two methods than can be used to minimize heaving. One consists of replacing the silt surrounding the footing with granular material. The second involves coating the foundation components with low temperature grease and wrapping with plastic film to prevent leaching. Both of these methods overcome the adhesion between the footing and the surrounding soil.

B   Manitoba Hydro Practice (McLure 1967)

(i) SELF-SUPPORTING (SUSPENSION) TOWERS

(a) *Peat depths greater than 8 feet (2.44 m)*   For depths of peat greater than 8 feet and where driven piles are inadequate owing to the presence of hardpan or bedrock at a depth insufficient to provide uplift resistance and/or lateral stability, islands of granular material are built. The peat is excavated during the winter and the excavation backfilled to the grillage level with coarse granular material. Compaction is carried out as required to employ the passive resistance offered by the bounding peat walls. After setting the standard steel grillage, the backfilling operation is continued with granular material.

The economy of this footing depends upon the availability of backfill material and the strength of the peat. It is necessary to observe the results, for the first few years at least, because of the tendency of the granular material to slump. Towers using this foundation treatment have remained essentially plumb for many years.

(b) *Shallow peat underlain by quicking sand*   In areas of shallow peat with the water table at the surface and underlain by quicking sand, steel H piles are driven to refusal. At each tower corner two piles are driven to depths of up to 40 feet (12.2 m). A template is used during driving to ensure that the piles are accurately located and is then left in place to provide lateral stability. Performance has been satisfactory and there has been no need to readjust the towers.

(c) *Depth of peat less than depth to underside of grillage*   In areas where the depth of the peat is less than the depth to the underside of the grillage, and where it is underlain by glacial till or clay (either of which provides excellent soil bearing capacity), the treatment illustrated by Figure 7.6 is called for. There are two alternative approaches which depend upon the economics of the situation. The adoption of a particular alternative should be preceded by a field test to demonstrate the degree of reliability provided.

Alternative I   All backfill material is placed in 6-inch (15.2-cm) lifts and compacted in accordance with the following requirements:

1. Where the water table does not occur within the footing depth and the installation is in till or clay, the first 3 feet (0.91 m) of backfill should be uniformly compacted to 95 per cent Proctor density. This plug should extend a full 3 feet

(0.91 m) above the top of the footing timbers. The remainder of the backfill and mounding should be compacted to 80–85 per cent Proctor density.

2. Where the water table occurs within the footing depth and the backfill material is till, the first 3 feet (0.91 m) of backfill over the footing timbers should be compacted to 95 per cent Proctor density and the remaining fill and mounding to 90 per cent Proctor density. If the backfill material is clay, the backfill for the entire footing depth and the mounding should be compacted to 90 per cent Proctor density.

3. Where suitable till or clay backfill is unavailable or uneconomical, granular backfill may be permitted, at the discretion of the engineer, provided that it can be compacted to 127 pounds per cubic foot (2035 kg per cu m) and that it contains little or no organic material.

4. Where peat occurs over the clay or till, the backfill should be built up as indicated in Figure 7.6.

Alternative II   Backfill (clay or till) should be compacted as follows:

1. The lower 3 feet (0.91 m) should be placed in 6-inch (15.2-cm) lifts and compacted by four complete passes of the compaction tools.

2. The balance of the backfill, including the mounding, should be placed in 12-inch (30.5-cm) lifts and compacted with two complete passes of the compaction tools. Granular materials may be used for backfill material. If peat occurs, fill should be replaced in accordance with Figure 7.6.

In practice, either of the above treatments may be expensive because of the inaccessibility of the sites, the high water table and the problems associated with

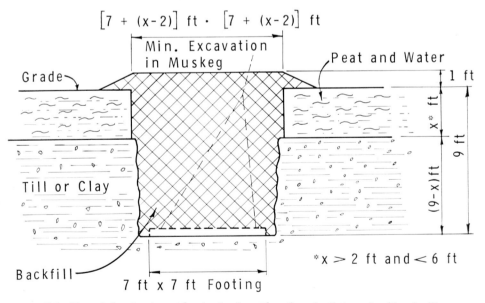

FIGURE 7.6   Foundation treatment for depth of peat less than depth to underside of grillage

excluding water, the unavailability of suitable backfill material, and the effort required to compact the backfill. The resulting foundation generally provides a greater than needed factor of safety as determined by tests.

(d) *Miscellaneous muskeg and high water table conditions* In an attempt to avoid the high costs associated with muskeg and high water table, the treatments indicated in Table 7.1 and Figures 7.7–7.12 have been used. These sketches are more or less self-explanatory and represent a variety of conditions encountered in the construction of one particular transmission line which followed a sand esker that rose a few feet above the muskeg on either side. It will be noted that timber grillages were used instead of the usual steel grillage. Frequent resort was made to a plank-type raft connected by steel channels to the stub leg. This grillage proved much more economical than one composed entirely of steel, especially where a large bearing area was required.

This type of treatment is adversely affected by either settlement of the grillage or seasonal loosening of the anchors due to frost heave. If the underlying soil is thought to be frost susceptible, as is the case with the finer permeable silty clays,

TABLE 7.1
Types of anchors used for various soil conditions

Description of conditions	Water table, peat or overburden depth	Backfill	Footing type
Anchors in rock	0–2 ft to sound rock	Replace original material to grade	Fig. 7.7
Anchors in rock	2–7 ft to sound rock	Replace original material to grade	Fig. 7.8
Sand below surface Depth of sand adequate for screw anchor installation	Water table 0 to 1 ft above sand surface	Replace sand to original level	Fig. 7.9
Sand below surface Depth of sand adequate for screw anchor installation	Water table 1 to 3 ft above sand surface	Replace sand to original level	Fig. 7.10
Depth of clay adequate for screw anchor installation	0 to 6 ft above clay surface	Replace original material to grade	Fig. 7.11
Depth of clay or sand *not* adequate for screw anchor installation Clay or sand overlain by 0–3 ft of peat	Water table at or near surface	Dumped sand or clay to surface	Fig. 7.12

the grillage is lowered below the frost line. Thus far, experience in sandy areas indicates that average settlement and/or seasonal heaving has resulted in a loosening of the nuts on the anchor rods of about ¼ inch (6.3 mm).

Varying soil conditions and the inability, at the present stage of technology, to predict in advance what penetration and holding power would be provided by screw anchors, made it necessary to provide for two, four, or six screw anchors at

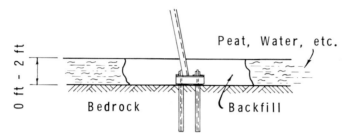

FIGURE 7.7   Anchors in rock underlying shallow peat

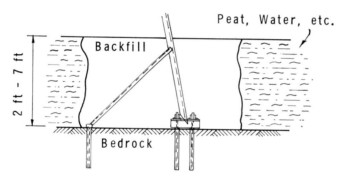

FIGURE 7.8   Anchors in rock underlying deep peat

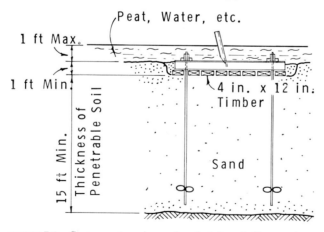

FIGURE 7.9   Screw anchors in sand underlying shallow peat

each tower leg. Small islands of dense sand or even frozen sand, which defied penetration, were encountered on the transmission line route mentioned above. Future developments will probably make much greater use of grouted anchors, in preference to the screw anchors used at present.

(ii) ANGLE TOWERS   One type of footing utilized for angle towers is illustrated by Figure 7.13.

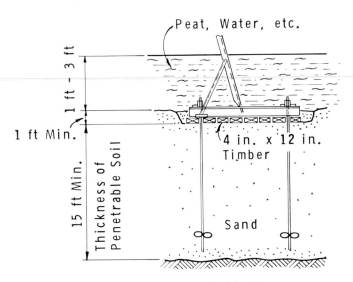

FIGURE 7.10   Screw anchors in sand underlying deep peat

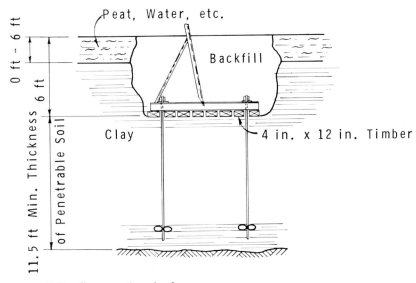

FIGURE 7.11   Screw anchors in clay

(iii) GUYED TOWERS   There appears to be some advantage, from the point of view of foundations, in the use of guyed rather than self-supporting towers in muskeg areas. Experience with guyed towers has been gained primarily in the construction of a transmission line in permafrost areas, but a considerable portion of the route traversed organic terrain. The types of anchors used under various muskeg conditions are shown in Figures 7.14–7.17.

In the initial stages of construction of the above transmission line, the more or less standard screw anchor was used, as shown in Figures 7.14 and 7.15. Field research was undertaken, however, into the use of a grouted anchor in tills, clays, silts, and even sands. Research results were sufficiently promising to permit the use of a grouted anchor throughout most of the line. The grouted anchor has the advantage that only one type of equipment (i.e. the overburden drilling equipment) and one type of anchor need be taken to each tower site. Reinforcing bars threaded at one end, cut to length in the field as required, constituted the anchor. Material brought up by the drill indicated to the operator the depth of anchoring required. The size of hole was increased to match the guy reaction.

The foundation under the central mast of the tower needs only to resist vertically downward applied loads. Problems associated with high water level are minimized and no compactive effort is required with the backfilling operation.

7.5   PIPELINES

Overland transportation of oil and gas has made necessary the construction of a network of pipelines in Canada to provide the most economic means of moving large volumes over long distances. These pipelines cross many types of terrain,

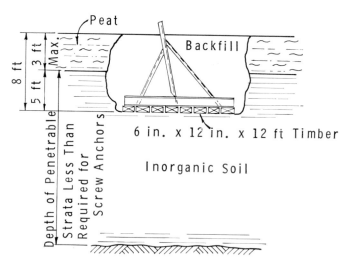

FIGURE 7.12   Raft foundation

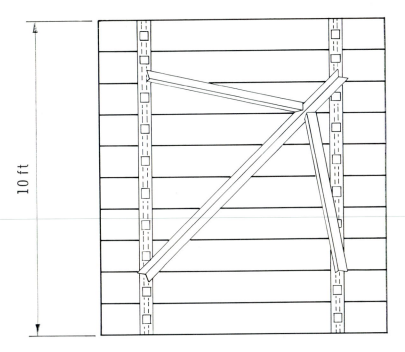

10 ft

PLAN VIEW

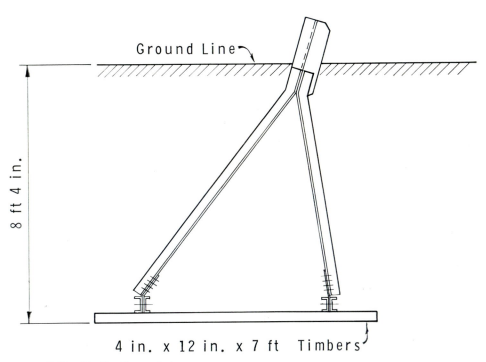

Ground Line

8 ft 4 in.

4 in. x 12 in. x 7 ft Timbers

FIGURE 7.13   Footing for angle towers

including muskeg (see Fig. 1.9). In fact, in many cases the oilfields are located in muskeg areas and consequently the gathering of oil and gas has raised difficult engineering problems.

Pipelining, like other types of construction, is preferably performed in warm weather. Where the bulk of the work is on mineral soil, there is usually a definite economic advantage in working in the warm and drier conditions of summer and fall. The most efficient performance will be obtained from both equipment and personnel under these conditions. When muskeg is encountered, it is usually

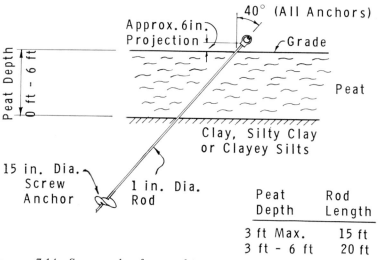

FIGURE 7.14   Screw anchor for guyed towers

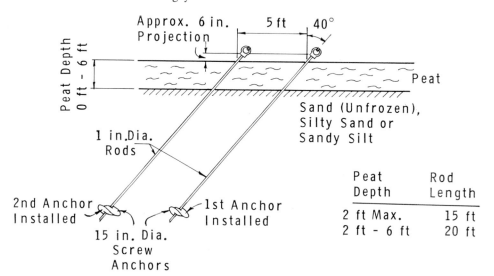

FIGURE 7.15   Screw anchors for guyed towers

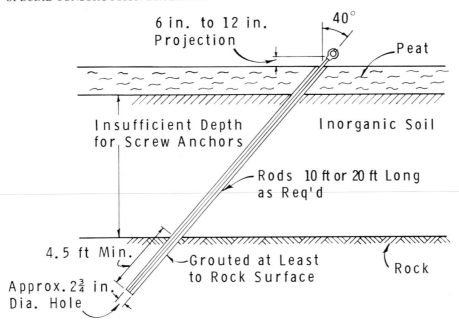

FIGURE 7.16   Grouted guy anchor in rock

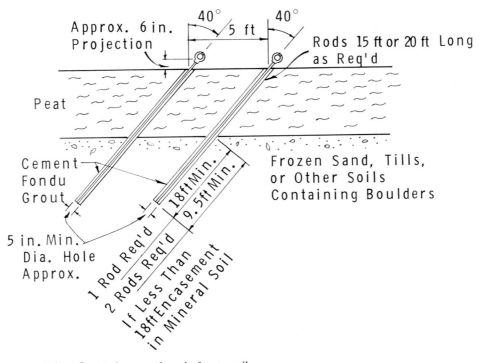

FIGURE 7.17   Grouted guy anchors in frozen soil

possible to "power" a way through it, although this may be costly and, as experience has shown, not necessarily the most desirable method.

Where the muskeg is patchy, the line relatively short, or construction timing dictates, warm weather construction over muskeg can be undertaken. First, the right-of-way is cleared through the muskeg areas, most often by hand, although light tracked vehicles with winches are a help. Usable timber is placed beside the ditch centre line as corduroy (Fig. 7.18). If necessary, additional poles are hauled in to completely corduroy the road area beside the ditch line where ditching, stringing, and laying equipment must travel. Light tracked vehicles for muskeg are used for emergency and the more difficult supply jobs. Ditching is carried out by using backhoes or shovels moving along the "road." In very wet muskeg, the trench sides are sloped flat to reduce sloughing; the extent of ditching ahead is held to a minimum for the same reason. Hauling, stringing, welding, coating, and lowering are done largely with conventional equipment and techniques, although modifications, such as load sizes, are made to suit conditions. Almost all pipelines in muskeg are weighted, as for river crossings, to create negative buoyancy, using concrete saddle or bolt-on weights (Figs. 7.19, 7.20, 7.21). Where the muskeg is underlain by solid mineral soil, hold-down clamps with earth anchors may be used to secure the pipe (Fig. 7.22). Particularly where the muskeg occurs in patches, long lengths of pipe may be strung, welded, and coated on solid terrain and pulled into place in the ditch, which has been made by a backhoe, on a "sled"

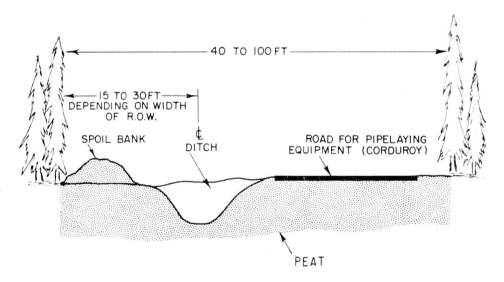

FIGURE 7.18 Pipelining in muskeg in summer. Depending on factors pipeline R.O.W. may be 40, 50, 60 up to 100 feet. Factors – terrain, land cost and availability, future looping and expansion plans. Centre line of ditch may vary from location illustrated in sketch above, where special conditions exist

pulled across the muskeg. Tracked vehicles will place saddle weights on the pipe as required.

Where most of the pipeline – or extensive sections at least – is to be built across muskeg and there is some freedom in the timing of the job, it has been found better from both an economic and technical standpoint to work in the winter.

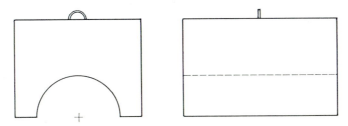

FIGURE 7.19   Concrete saddle weight

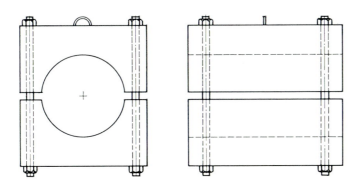

CONCRETE BOLT-ON (CLAMP) WEIGHT

FIGURE 7.20   Pipe weighting

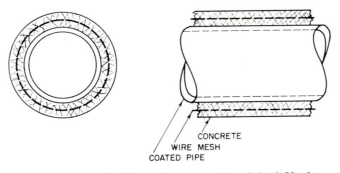

CONCRETE
WIRE MESH
COATED PIPE

FIGURE 7.21   Gunnite (sprayed on concrete) weighted. Used mainly in water crossings or extremely wet terrain

A detailed account of the principles of construction on frozen muskeg is given in Section 7.6; in the following paragraphs, however, the approach to winter construction is described as it applies specifically to pipelines.

The right-of-way is cleared of all brush and debris, preferably in the fall or the early part of winter, although this work may proceed well into the winter. As before, light tracked vehicles are useful for clearing, although it remains largely a manual operation. After the muskeg has frozen sufficiently, a light bulldozer pushes the snow off the area to be used as a road and piles it over the ditch line. Some placing of corduroy on very wet muskeg may be beneficial during this phase of the operation. This snowploughing allows the frost to penetrate deep into the road area and it effectively grades the right-of-way for later equipment movement, while the snow cover prevents the ditch line from freezing (Fig. 7.23). As soon as the roadway is capable of supporting the heavy pipeline equipment, construction is started. The snow is cleared from the ditch-line area and conventional trenching machines will easily handle the light frost penetration that has occurred under the snow. Because trenching machines effectively straddle the ditch line, special precautions may be necessary. These include extra wide tracks, temporary mats for the tracks, or lightening (i.e., bearing part of the weight of the trencher) by large side boom tractors moving along the road. When a trencher cannot be used, the ditch is dug

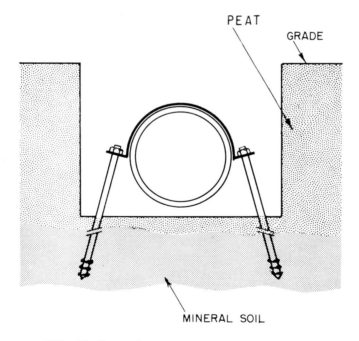

FIGURE 7.22   Pipeline anchor

by a backhoe moving along the frozen road. The ditch will stand up well and very few problems are encountered since all the equipment can work on and from the frozen road.

Cold weather construction techniques (operable except in extremely cold and stormy weather) have been developed: for example, preheating the pipe for welding; using two welders for each weld pass (in medium and large diameter pipes); keeping weld passes consecutive to minimize cooling in between; controlling post-weld cooling by wrapping pipe with insulating blankets; warming pipe for coating or using yard-coated pipe so that only the joints require coating in the field, etc. For cold weather construction, it is desirable to keep the pipeline spread in tight formation so that all operations from stripping snow to clean-up are less spread out. This reduces problems such as the trench filling with water which then freezes, sloughing of ditch sides, and freezing of the spoil bank (ditch excavation material) which increases backfill and clean-up work. Although conditions vary from year to year, it has been found that the last half of January and the months of February and March are ideal for pipelining in muskeg.

Techniques have been developed, particularly in Great Britain, whereby cables and pipelines are ploughed into muskegs or softer soils. A patented plough is used in which the depth of the ditch is controlled by the approach angle of the cutting edge of the plough. Two methods are employed: in one the cable or small pipe is on a reel which moves on or with the plough; in the other, the pipeline is laid out on solid ground and then pulled by winches into the ditch which is opened by the

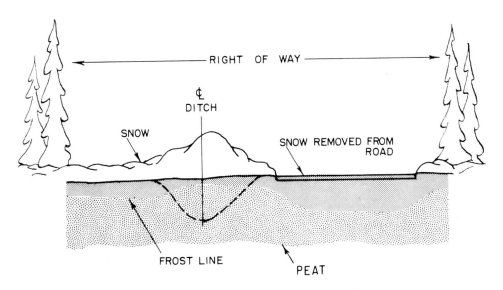

FIGURE 7.23   Pipelining in muskeg in winter

plough in up to ½-mile (0.8-km) segments. This system has been used on pipe up to 8 inches (20.3 cm) in diameter and, within certain limits, is economical and relatively fast. It has certain disadvantages. For example, the coating is frequently damaged unless special additional protection is added; it is suitable mainly for flat terrain since the pipe cannot readily be bent to land contours; it does not easily provide for weighting the pipe and – unless extensive use could be made of it on a job – setting up such an operation may be less desirable than the pulling method referred to in summer construction. Its use for small sizes appears to hold good promise.

Muskeg classification systems or other evaluations can frequently provide valuable assistance in pipeline route selection. Owing to the cost of the pipe, coating, right-of-way, etc., the shortest, most direct route is the first choice regardless of soil or topography. The larger the pipe diameter, the more this straight-line route selection applies, in view of the relation of the value of the pipe to other costs. Deviations from the straight line, however, due to unfavourable terrain, access during and after construction, etc. will often be found desirable both for economic and construction reasons. Here, muskeg classification from aerial photographs can be of value. It is also useful to the contractor as a guide in pricing the line, predicting trouble spots, and planning the work.

7.6   CONSTRUCTION OVER FROZEN MUSKEG

Construction over frozen muskeg presents a special problem to the engineer. This problem is related almost entirely to the possibility that the frozen terrain will thaw and revert to its normal organic state. For a discussion of this construction problem, frozen muskeg is considered in the three categories set forth in the following paragraphs. Although road construction over frozen muskeg is the chief concern here, the principles outlined are equally applicable to other types of embankment construction.

(1)   *Categories of Frozen Muskeg*

A   Category 1: Seasonally Frozen Muskeg
This type is found south of the permafrost line anywhere in Canada and, as the name implies, the frozen zone is entirely seasonal. The depth of frost penetration or the thickness of the frozen layer is highly variable depending upon climatic factors, such as temperature and wind, and on the snow cover. In the more southerly areas of the country, it may be as little as a few inches, and hence is unsafe for heavy equipment. The depth of the muskeg and the Radforth classification (see Chapter 2) vary widely and each must be determined for the specific project under consideration. It has been observed that for muskegs lacking tree growth or for areas of larger muskegs free of trees (that is, the predominantly EI type), there are

often pools of water with a diameter of about 10 feet (3.05 m). Under a snow cover, the ice forming on these pools may be very thin, and can easily be broken. Recovery of equipment under these conditions is difficult and costly. The presence of these pools, and the hazard they present, may be evaluated from summer airphotos.

B   Category 2: Muskeg in Continuous Permafrost Areas
Experience has shown that muskegs of permafrost areas are generally relatively shallow, their depth seldom exceeding a few feet. Geologically, this condition would appear to be widespread, especially for glaciated areas. Many of the glaciated valleys of the Yukon Territory appear to be floored with a gravelly till. There is, however, much geological evidence of transient Pleistocene lakes in areas now covered with muskeg. These two factors lead to the generality that muskegs in the permafrost areas of the Yukon are underlain by a gravelly till or silt. Farther to the east in the Canadian Shield area, one might anticipate rock underlying the muskeg as the organic growth in filled shallow rock basins. In general, the muskeg cover is of the EI type with a few free-growing bushes of more than 2 feet (0.61 m) or so in height. One rather obvious exception to the Radforth classification suggested is the so-called "drunken forest." It is believed that this phenomenon is more properly thermokarst topography (see Glossary) than muskeg and, in any event, is to be avoided where possible.

C   Category 3: Muskeg in Discontinuous Permafrost Areas
This category presents the most serious problems to road and other construction. Its geographic distribution has been outlined (Brown 1965; Brown and Johnston 1964). It is difficult to differentiate between seasonally frozen muskeg and permafrost; it is equally difficult to determine the limits of the permafrost areas that may be present. This category is transitional between the first two; hence the vegetative cover and the underlying inorganic soil vary widely.

(2)   *Properties and Preliminary Studies*
In any construction involving muskeg, the thickness of organic material, its classification, and the type of underlying inorganic soil must be determined. The procedures for such investigations have already been outlined (see Chapter 5). In addition to these standard investigations, however, information must be obtained concerning the frozen zones. In seasonally frozen muskegs, the thickness of the frozen layer should be determined. The critical areas of a muskeg in which such thicknesses should be determined are the open areas where tree and shrub growth is at a minimum. Usually 10–12 inches (25.4–30.5 cm) of frozen peat will support most heavy equipment and this thickness can be increased by stripping off the snow cover to allow deeper frost penetration.

The depth of the muskeg in permafrost areas, as well as the thickness of the active zone and the type of the underlying inorganic soil, should also be established. Areas of ice wedges and the ice content of the frozen soil should be noted and plotted on the preliminary plans.

(3) *Design and Construction*

A   Seasonally Frozen Muskeg

Construction on seasonally frozen muskeg involves essentially the same design criteria considered in previous chapters which deal with unfrozen muskeg, including subsurface investigation and provision for drainage. The use of design criteria for unfrozen muskeg is based on the fact that thawing of both the seasonally frozen zone and areas of permafrost will occur.

Frozen muskeg offers a solution to the problem of access to organic terrain if there is a foot or so of frost penetration under the snow cover. The tractor train and the dog team are well-known examples of the mobility offered by frozen ground. The price paid for this advantage is winter work. Cold weather clearing of trees, shrubs, etc. to the general muskeg level is usually easy since at this time organic growth snaps off readily at ground level.

Material handling during cold weather requires that any fill built on muskeg must be basically a cohesionless material, a sand, gravel, or quarried rock, or that fill operations be carried out on a 24-hour round-the-clock basis. It is debatable whether the fill materials need be of high quality since this factor is often controlled by local borrow materials.

Most muskegs over which a road or other embankment will be built have a surface mat of living organic material. Care must be taken that coarse fill material does not punch through this surface material when thawing occurs. The punching of a large rock, for example, leads to a local subsidence of the fill which will quickly be made worse by traffic. Of more concern is the fact that the organic material makes its way into the fill, resulting in a deterioration of its strength. The soft spot enlarges and a rotational failure often ensues. The integrity of the surface material can be preserved and the punching of large aggregate can be prevented by first spreading a sand or sandy gravel material about 12 inches (30.5 cm) thick, then building the fill on this granular material. The rocks in the subsequent fill should not exceed 1 or 2 feet (0.3 or 0.61 m) in diameter, depending on the fill height. The nature of the local materials available for construction will influence the size. If a free-draining material is used in the fill, the presence of snow, if not excessive, will do no harm and, in general, may be ignored.

Under the influence of the fill, the time rate of thawing of the frozen material is variable but may take several years. Thawing will inevitably be accompanied by settlement of the fill. The amount and distribution of this settlement depends on the vagaries of the particular muskeg. It is not predictable in detail, therefore, unless subsurface data are available and the settlement characteristics are known. Observations of settlement are of considerable value in keeping track of the progress of the settlement, but such records do not necessarily involve elaborate or expensive equipment. Simple settlement stations and careful levels will provide adequate data. Advice on field installations and observations is given in Section 6.5(4).

There are, of course, alternatives to the suggestions in the preceding paragraphs. The amount of winter work can be reduced by building only a portion of the fill during cold weather and using this as an access mat during the following summer season. A surcharge may be placed on the fill and allowed to play its part during the cold weather and early spring. The surcharge may be used as a stockpile for adjacent fills to be built in subsequent operations.

B   Muskeg of Permafrost Areas

Roads and other embankments over muskegs of permafrost areas can be conveniently considered under two subgroups: first, muskegs underlain by rock or gravel; second, those underlain by silts or materials with a high ice content. Given an adequate foundation material and a shallow muskeg, a fill may be placed as for a normal operation, with removal of the organic material being a matter of choice for the particular site. For a first-class road, for instance, its removal is recommended. This is accomplished by removing layers as thawing progresses. In those cases where the organic material is not removed, it is only necessary to clear the area of free growth or for higher fills to flatten the growth.

In the second group (that is, silts or materials having a high ice content), the design is usually based on the premise of preserving the frozen state. Permafrost is particularly sensitive to thermal changes and practically any man-made changes upset the dynamic equilibrium established by nature (Brown and Johnston 1964). If thawing is allowed to take place, the water cannot drain away and a pocket of viscous slurry forms. The fill settles into this pocket and becomes soft because of the water or contamination by the slurry. In either event a costly repair problem ensues.

To maintain the permafrost, a minimum thickness of fill of the order of 6 feet (1.83 m) should be planned and berms 6–8 feet (1.83–2.44 m) wide and 2–3 feet (0.61–0.91 m) thick should be considered along the sides of the fill. In view of the sensitivity of the permafrost to environmental changes, an absolute minimum of disturbance should precede placing of the fill. The free-growing material should be cleared or flattened, but this phase should barely keep ahead of the fill.

Hemstock (1967) refers to a permafrost road technique developed in Western Siberia by the Russian Oil Ministry to improve access to oil drilling sites during the summer. It is somewhat of a misnomer since there is little, if any, natural permafrost in the area. By their method, however, permafrost is formed and then protected throughout the summer to provide a solid base for drilling site access roads. The method is used as far south as the 60th parallel. During the winter, the proposed road right-of-way is kept clear of snow and any insulating moss is removed. Frost penetration of about 1.5 metres (5 feet) results. Just before break-up, the right-of-way is covered with peat from the ditches and right-of-way, over which is spread a layer of sand. This preserves the ice, thus making a solid base for summer travel. Large ditches are provided to help with drainage. The mixture of sand and peat is considered to be better than wood chips in that it stands up better to traffic.

These roads are fairly expensive ($17,000 per mile) but are used for access to drilling islands where there is good economic incentive. Some 25 kilometres (15 miles) were built in this manner during the winter of 1967–68.

c   Muskeg in Discontinuous Permafrost Areas

The third case comprising muskegs in the areas of discontinuous permafrost (Brown 1965; Brown and Johnston 1964) presents the most serious problems. As stated earlier, it is difficult to differentiate between seasonally frozen muskeg and permafrost; it is equally difficult to determine the limits of the permafrost areas that may be present. In the geographic localities in which such problems are encountered, the peat may vary considerably in depth, the permafrost often contains lenses of clear blue ice and is very sensitive to any environmental changes. On this basis it appears virtually impossible to preserve the frozen state; therefore, design procedures must be similar to those drawn up for the unfrozen state.

If feasible, the presence of the ice lenses and areas of permafrost should be determined and the areas so delineated avoided. If it is not possible to avoid them or it is not desirable for other reasons, then the strong probability of inordinate settlements and high maintenance must be accepted. It is advantageous to clear the proposed or tentative right-of-way in the late winter of one season and allow the area to lie dormant over the ensuing summer. In the early fall, surface and subsurface investigations may be undertaken, the latter combining boreholes and simple probes. The information obtained will allow more detailed planning such as choice of centre line, height of fills, and location of surface drainage.

(4)   Maintenance

It must be clearly recognized and accepted that maintenance subsequent to construction will be very high for roads and other embankments over seasonally frozen muskeg, and in areas of discontinuous permafrost, for the first 3–5 years of the life of the structure. Ideally, much of this maintenance effort should be considered in the design stages. The maintenance of embankments is largely directed towards bringing them back to grade elevation as settlement occurs. A stockpile of material adjacent to muskeg sections is of particular help in reducing the work of filling in settled areas. The maintenance effort diminishes in succeeding years until a new dynamic equilibrium is established.

(5)   General Considerations

The following paragraphs deal with general points that have a bearing on the problems of roads and other embankments over frozen muskeg. The order of presentation is one of convenience and does not suggest degree of importance.

If long-range planning is possible, it is highly desirable, after the late winter clearing, to establish drainage ditches as early as conditions allow. The effect of these ditches operating over the summer months often eases the problems of design, access, and construction.

A very limited experience with explosives in ditching muskeg in the frozen state indicates that it may have some merit. If the peat is not too fibrous to preclude the drilling or steaming of holes, then a single row placed at an angle to the horizontal will create a ditch. The spacing and charge are matters for experiment, but stemming is a necessity. Otherwise, the explosive energy is dissipated by coming up the loading hole and creating large in-place blocks of frozen material. It is for the same reason that the holes must be slanted.

Muskegs located in valleys very often have a gentle slope towards the valley centre. This gives rise to a flow of water approximately perpendicular to, as well as along, the valley axis. A road, therefore, almost invariably interferes with the natural drainage regime. In permafrost areas, the seepage zone is restricted to the upper active zone or to unfrozen layers (taliks) (Muller 1947); the road fill may cause these seepage zones to freeze off. The seeping water continues to flow to the frozen area, however, and large masses of ice build up. These ice masses are referred to as "icings" (Muller 1947; Thomson 1963). They can grow to the point where the ice encroaches on the road surface, thus creating a serious driving hazard. These phenomena are essentially a drainage problem and adequate ditches and culverts must be provided. The road surface should be not less than about 6 feet (1.83 m) above the adjacent muskeg level to maintain the original permafrost, to aid in snow clearance, and to prevent encroachment of icings on the road surface.

The problems involved in paving roads over frozen muskeg appear to be in a speculative stage. One can only make such suggestions as delaying paving until some degree of fill stability has been achieved and placing a light-coloured chip coat on the road surface to retard heat absorption by the fill. It may be desirable to increase the fill thickness as another means of preserving the underlying permafrost.

The problem of access route location in northern areas has different aspects than in populated areas (Savage 1965). The location engineer has a much greater latitude since land purchase is rarely a consideration. It is possible, therefore, that the location of a road, railway, etc. can be chosen to avoid many difficult problems. The use of airphotos in the planning phase has much merit (Mollard 1961; Mollard and Pihlainen 1963) and is of particular assistance in the location of borrow-pits or general construction materials in permafrost areas.

Maintenance operations are a costly and often tedious chore. The adoption of any design features that can be incorporated into the original construction to reduce subsequent maintenance are strongly recommended. These features include final selection of the road centre line, adequate fill height, side berms, drainage, and, where possible, the selection of quality fill material. A factor in some maintenance repair problems is that the heavy equipment present for the construction is not available for repair work.

7.7 CORROSION IN MUSKEG

Reference is made in Section 4.2(8) to the acidic reaction of most peats, which is due to the presence of carbon dioxide and humic acid arising out of the decay of the organic matter. Concrete and metal structures in such an environment, therefore, are subject to corrosive attack and precautionary measures may be necessary.

The run-off waters from muskeg areas have much the same general aggressive properties as the peat and may have a deleterious affect on any structure with which they come in contact. Although muskeg waters are usually acidic, alkaline waters do occur and may also be corrosive. Running water is potentially more harmful than stagnant water. Furthermore, attack at a constantly changing water level is stronger than attack on a completely immersed structure. Frequently, the run-off water from muskeg areas is brownish in colour, but the intensity of the colour is not a good indication of its potential aggressiveness. The principles of corrosion in muskeg areas are described in the literature (Brawner and Darby 1962; MacFarlane 1965; Mainland 1962) and will be referred to only briefly here.

There is very little quantitative information of value to the design engineer concerning the severity of deterioration or corrosion of structures in muskeg. The small amount of information that is available, however, indicates that the type and rate of attack does follow a predictable pattern (Mainland 1962).

The most common structures in muskeg areas are culverts. Table 7.2 presents an assessment of the merits of various types of culverts relative to the corrosion problem (Brawner and Darby 1962).

(1) *Concrete*

Deterioration of concrete by the action of aggressive waters is a chemical action which is potentially serious if the water is able to percolate through the mass. When this water is very pure or contains free carbon dioxide and acids, it will attack set cements. The degree of acidity, however, does not give a simple measure of the aggressive action, although it does bear some relation to it. A chemical analysis of the water, especially with regard to free carbon dioxide, is helpful in predicting aggressiveness.

The temporary hardness of peatland waters, together with their pH, may also be used to predict aggressiveness to concrete. As a broad guide, a water with a temporary hardness of 10–20 parts per million (p.p.m.) will not have a marked solvent action unless the free carbon dioxide is above 10 p.p.m. (Lea 1956). For a temporary hardness of 5 p.p.m. the corresponding critical free carbon dioxide is 5 p.p.m., whereas water with a temporary hardness of less than 2.5 p.p.m. can be aggressive even if the free carbon dioxide is negligible. Water may be aggressive for pH values up to 7.0 or even 7.5 if the temporary hardness is less than 2.5 p.p.m., and at pH values of up to 6.0 or 6.5 for higher values of temporary hardness.

TABLE 7.2
Comparison of corrosion or deterioration of culverts in muskeg areas
(after Brawner and Darby 1962)

Culvert type	Corrosion or deterioration
Concrete	Subject to acid and sulphate attack. Rate of attack variable; specify high density concrete, air entrainment
Galvanized metal	Subject to corrosion. Rate of attack variable but often rapid. Use of cathodic protection beneficial but not always economical
Bituminous coated metal	Corrosion reduced by asphalt coating with rate of deterioration dependent on conditions of coating. Can consider cathodic protection
Aluminum	Ability to withstand corrosion is generally unknown. Requires further evaluation. Field test installations suggested
Untreated wood	Deteriorates rapidly above water
Creosoted wood stave	Very resistant to deterioration

Deterioration of concrete is also a function of the quality of the concrete since it depends to a large extent on the permeability of the cement paste. This, in turn, is related to the water–cement ratio as well as to the compaction of the mix (Powers *et al.* 1959). Thin-walled concrete structures such as drain tile and culverts are much more susceptible to aggressive action than are mass concrete structures. Also, fresh concrete is more vulnerable in the presence of aggressive water than is mature concrete.

In order to resist attack by aggressive waters, concrete must be of good quality, and dense. High-alumina cement and air-entraining agents provide further protection in extremely aggressive environments. Many surface treatments have a life considerably shorter than the design life of the structure and have not been too satisfactory over a long period of time (MacFarlane 1965).

(2)   *Metals*
Corrosion of metals in a muskeg environment is an electrochemical action. The type and rate of corrosion are functions of the properties of the metal as well as of the soil and water conditions. Factors such as physical properties of the soil, dissolved salts, pH, total acidity, resistivity, aeration, and presence of anaerobic bacteria influence corrosion. Methods for estimating soil corrosiveness will usually entail the measurement of one or several of these various factors.

Low electrical resistivity of the soil and relatively high quantities of soluble salts generally indicate high corrosivity. An acid soil (such as peat) would also be expected to be corrosive because of its tendency to prevent the formation of protective films. Oxygen, either from the atmosphere or from oxidizing salts or compounds, also tends to stimulate corrosion. Micro-organisms which occur under anaerobic conditions and which are associated with soil corrosion thrive in the absence of oxygen and in environments that are almost neutral or slightly acid, for example at pH 6–9. These bacteria are believed to occur frequently in peat soils.

A practical indicator of potential corrosiveness is the soil resistivity; variations of Table 7.3 are commonly employed (Mainland 1962). The resistivity depends upon the water content of the soil and the dissolved salts in the water. In natural peatland, the water content is usually high and the salt content relatively low; generally the resistivity will be over 1000 ohm-cm. Resistivity readings taken in peat in the vicinity of the proposed structure, together with an assessment of the overall drainage pattern, muskeg type, and other features, will indicate whether further information should be obtained. Economic considerations normally will dictate whether it is advisable to obtain such detailed information as pH, total acidity, soluble salts, available oxygen, presence of micro-organisms, etc.

TABLE 7.3
Relation of corrosion to soil resistivity
(after Mainland 1962)

Soil resistivity	Corrosion potential
0 – 1,000 ohm-cm	Severe
1,000– 10,000 ohm-cm	Moderate
10,000–100,000 ohm-cm	Very moderate
100,000 and over	Non-corrosive

When the building of a metallic structure in muskeg is under consideration, the following points should be kept in mind (Mainland 1962): (i) if a structure is entirely immersed in a large body of muskeg, corrosion should be fairly slow and will be of a uniform nature. (ii) If it is only partially immersed, some corrosion can be expected due to differential aeration cells. (iii) If sulphate reducing or other type of depolarizing bacteria are present in the peat, the corrosion rate in the metal will be appreciably increased. (iv) If the structure is a long one, such as a pipeline, and is buried in peat and in adjacent inorganic soil, very serious corrosion can normally be anticipated near the edge of the muskeg.

Measures for countering the effect of an aggressive environment on metals include coating and cathodic protection. One method is to coat the metal by galvanizing. A zinc coating insulates the steel from the electrolyte and provides some anodic protection when small areas of steel become exposed. In highly acid waters the zinc coating is likely to last from 2 to 5 years. Additional protection can be provided by coating the galvanized pipe with a relatively thick layer of asphalt. The asphalt is resistant to acid attack and also reasonably impermeable to water. The major problem is to ensure good adhesion between the asphalt and the zinc coating. In some instances, asbestos fibres are embedded in the zinc at the time of galvanizing to provide an anchor for the asphalt coating.

Where very large and expensive installations are involved, cathodic protection may be utilized. Sacrificial cathodic protection involves creating a corrosion cell whereby magnesium or zinc anodes are installed in the soil or water (electrolytes) and connected by metal paths to the structure (pipe, culvert, etc.), which then becomes the cathode or protected electrode of the system. In the case of culverts, because two electrolytes are involved – soil outside the culvert and water inside – two sets of anodes are required. One set is buried in the soil and the other is placed in the invert of the culvert.

Cathodic protection is normally utilized in conjunction with a protective coating of asphalt since the amount and cost of cathodic protection is greatly reduced with the use of such a coating (Brawner and Darby 1962).

(3) *Wood*

Wood is subject to attack by fungi and insects; in both cases the wood must be moist and a certain amount of air must be present (Brawner and Darby 1962). For culverts in particular, such conditions will prevail in at least some portion of the structure in most installations. Since the fungi and insects are dependent on the wood itself for sustenance, the wood may be "poisoned" by pressure treatment with creosote. Neither the wood itself nor the creosote is subject to attack from the acid conditions likely to prevail in most muskeg areas. Where wood structures are kept completely immersed, therefore, they will be preserved for a very long time.

7.8 DRAINAGE OF MUSKEG

(1) *Artificial Drainage*

Cuthbertson asserted that at best muskeg can be drained only imperfectly and then at surface level and with closely placed drains (*in* Johnston and Hills 1956). Certain drainage experiments performed north of Iroquois Falls in Northern Ontario by the Abitibi Power and Paper Company partly supported this view. Ditches 3.5 feet (1.07 m) or more in depth were imposed on a 60-acre (24.3-ha) tract of muskeg which was mostly wooded. The drainage effect was measured as a function of the increased growth rate of the trees distributed over the drainage grid. It was concluded that drainage was ineffective beyond 2–3 yards (1.83–2.75 m) from the edges of the ditches. The total amount of growth increase was insufficient to offset the cost of ditching. It would appear that in over 30 years, the duration of the experiment, there has been no improvement in this situation. Furthermore, trees planted on small non-forested areas of the tract did not establish themselves despite the attempts at drainage.

Examination of the tract shows that the ditches were placed in accordance with a geometric pattern and without regard to changes in the biological organization in the muskeg. Thus, the ditches did not conform to the natural drainage

FIGURE 7.24   Site of local perched water table near Parry Sound, Ont. A perched or false water table exists at the surface of the EI and FI beyond the major line of demarcation. In foreground FI, only the main (lower) water table exists

pattern. Some of the ditches appeared as open linear reservoirs and were not functioning as drains.

(2)   *Natural Drainage*
Drainability varies, depending primarily on the physiographic relationships of the muskeg with its physical environment, and secondarily on the type of peat and covering vegetation. It is difficult to separate the two conditions. In certain seasonal or transient circumstances there may be a low main water table and a local perched water table in muskeg where EI or EFI (see Chapter 2) prevails (Fig. 7.24). In aapa moor with residual ice and in raised bogs late in the summer (perhaps for the entire year), though the general water table is at a relatively low level, buried plateaus of ice or sometimes elevated islands of ice exist in the peat (Fig. 7.25). The water or ice in the false water table near the surface is thought to be the compound effect of some free and some capillary water. Lake (1961) has shown that, when a hole is dug in muskeg of this kind, free water will gradually seep into the cavity. It is also known that in hillside muskeg open water in natural ponds may persist at different levels down the slope (Fig. 7.26).

Assuming that the water is free to flow, experience indicates that in late summer or when water tables are low, the driest peats are in areas where persistent or

FIGURE 7.25   Elevated ice mass in peat above water table

frequent drying winds occur, as in Newfoundland (Fig. 7.27), or in Northern Canada on the Hudson Bay lowlands (Fig. 7.28). The effect however is local; a high incidence of imponding characterizes such areas (Fig. 7.29). The persistently wettest muskeg is at the level of the main water table. The commonest cover formulae conveying this environment are FI and FIE. If the water table is lowered in these cases, the peat will usually drain readily without leaving a residual perched effect.

Depending upon the season of the year and on the conditions of rainfall, all muskeg types show a range of tolerance to the presence of free water. Free water may be at the surface and visible in the cover or may not be exposed until the surface of the muskeg is disturbed mechanically. In either case, and despite the range in the amount of free water, experience indicates that the differential in relative dryness is in accordance with the following sequence of cover formulae, with the driest on the left:

HE, EH, EFH, FEH, AEH, BEH, AEI, BEI, ADE, BDE, BFD, BDF, DEI, EI,

EFI, DF, DFI, FEI, FI.

Where there are wide differences in cover and peat type in a given tract, it seems reasonable to attempt drainage from drier to wetter areas. Usually the wettest conditions characterized by FI are associated somewhere on their margin with DFI or DI. The gradient is then invariably towards DFI and its course follows the

FIGURE 7.26   Local water tables in sloping muskeg showing difference in elevation (near Prince Rupert, BC)

FIGURE 7.27   Muskeg subject to effect of drying winds near Trans-Canada Highway, east Newfoundland

FIGURE 7.28   Windswept aapa showing peat plateaus in permafrost region near Fort
Churchill, Man.

FIGURE 7.29   Aerial oblique showing pond effect in windswept muskeg south of Fort
Churchill, Man.

FIGURE 7.30   Linear drainage (left to right) through muskeg characterizing water regulations for DFI cover

direction of DFI (Fig. 7.30), which is for the most part linearly oriented though sometimes tortuous. It is wiser, therefore, to exploit the natural gradient than to impose on the terrain a symmetrical geometrical system that might work against the natural trend or be ineffective. It is only when a single muskeg type exists that a geometrical pattern of artificial gradients may be imposed with relative success, but this circumstance is unusual.

The drainage relations discussed so far are for local conditions. Their effectiveness may be enhanced by using the wider, primary geomorphic setting in which one or several local conditions may be found. In unconfined muskeg, the major and ultimate course of drainage should relate to the major geomorphic conformation. The collected free water should be led in the direction of the base of the major ultimate drainage gradient.

Confined muskegs are usually associated with the lagg surrounding the entire peat deposit. Often there is directional flow within the lagg and this leads to an outlet which may be deepened if drainage is contemplated. Elaborate precautions should be taken when building roads across confined muskeg to position culverts in the lagg area to prevent ponding at either side of the road embankment.

(3) *Drainage Aids*

Pihlainen (1963) points out that "even though the fundamental principles of mus-keg and peat drainage are not even partly understood, a considerable amount of empirical knowledge on the reclamation of peat through drainage is available." In terms of method, Cuthbertson (1959) indicates two kinds of drainage, one of which deals with underground installation of tiles, timberbox sections, mole drains, or perforated plastic pipe. The second approach is known as the open drainage system, in which open ditches are imposed on the land with the object of enhanc-ing any natural drainage trend.

Underground drainage has the advantage of leaving the surface of the muskeg unscarred for agricultural use. If trench draining is employed, care must be taken not to interfere with the requisite conditions for the operation intended to utilize the land. In this connection, there are certain limitations. For sloping muskeg, the ditches must conform to those directions which will catch run-off roughly at right angles to the slope of the natural gradient. Machines are now available which can cut drainage ditches at a rate of 1–3 miles (1.61–4.83 km) per hour for a 3 foot (0.91 m) deep ditch. Major ditches and off-takes are best constructed with the aid of a dragline which can handle a variety of conditions effectively (Fig. 7.31). Ayers (1964), in contemplating drainage for large reclamation areas, specifies primary, secondary, and tertiary drainage measures, after all attempts to control the water entering the muskeg basin have been met.

In the primary phase, open ditches are placed at the topographic lows at the suggested formula of 1 mile (1.61 km) of ditch per square mile (259 ha) of area as a minimum. If it is not possible to place such ditches at sufficiently low elevation and with appropriate grade to prevent flooding, then a pumping system will have to be installed at the lowest point of the drainage system in order to lift the water at appropriate capacity over the land mass or obstacle.

In the secondary drainage, shallow, 4-foot (1.22-m) ditches are placed in accordance with the programme for which the land is to be utilized (possibly a cropping plan). It is recommended that the ditches be spaced at 300-foot (91.5-m) intervals and that the land be graded to facilitate surface drainage in the direction of these ditches.

In tertiary drainage, an underground system of tile drainage is employed rather than open ditches. Cuthbertson (1959) suggests the possibility of using perforated plastic pipes which, though expensive, are easy to install and function effectively. It is recommended that the drainage be placed to a depth of 5–6 feet (1.53–1.83 m), care being taken not to place the tile or pipe in impermeable subsoil. Spacing of the drains is a matter of judgment, but it depends upon the properties of the particular peat type, the amount of residual water required, and the precipi-tation statistics.

In ditching, care must be taken not to block surface run-off by the spoil that comes up with the plough. Healy (1961) used a rotary ditcher by which the spoil is disseminated over a large area and drainage at the shoulder of the ditch is thus

FIGURE 7.31   Ditch excavation by dragline

unimpeded. With this device there is little possibility of clogging, a situation experienced with the kind of drainage plough he had used formerly.

In forested muskeg, drainage ditches are exceedingly difficult to plough out
and even the dragline method produces complications and bad sloughing from the
sides of the ditches. Berglund and Akelyon (1949) recommend the use of
dynamite. Charge holes are made along the line where the ditch is to run. The
charges are placed 8–16 inches (20.3–40.6 cm) apart and are exploded by
"quick-match and percussion cap." A chain reaction is set up from the successive
bursts and the charges act almost collectively even over a long stretch. For 900 m
(2950 feet) of ditching, 350 kg (772 pounds) of drainage dynamite were utilized
in the draining of 375 acres (152 ha) of muskeg in Sweden.

For hillside muskeg, care should be taken to keep the sides of the ditch as
smooth as possible; otherwise an irregularity will start erosion problems. To effect
reclamation of cover or change in the cover of the muskeg, ditching can be
shallow. Ditches 19 inches (48.3 cm) wide at the top, 8 inches (20.3 cm) at the
bottom, and 20 inches (50.8 cm) deep are ideal. Ditching of this type is being used
in Scotland and the Falkland Islands where recently 1000 miles (1610 km) of
ditching was done on one farm alone (Falkland Islands Agric. Dept. 1967).

The practice of ditching in muskeg is common in European countries. By 1940,
about 800,000 ha (3090 square miles) of muskeg had been reclaimed in Finland.

Following World War II, approximately 250,000 km (155,200 miles) of forest ditches had been dug and ditching proceeds on the basis of about 25,000 km (15,520 miles) per year, involving the drainage of approximately 120,000 ha (465 square miles) (Heikurainen 1968).It was indicated by P'yavchenko (1963) that drainage of muskeg for forestry purposes is being exploited in the USSR. There, the peat is blown out for a distance of 100–200 m (328–656 feet) and more with the aid of fused charges of "demolition explosives." This method has been demonstrated by the Forestry and Timber Institute of the Siberian Department of the Academy of Sciences in the USSR.

REFERENCES

AYERS, H. D. 1964. Engineering problems in the exploitation of organic terrain for agricultural reclamation. Proc. Ninth Muskeg Res. Conf., NRC, ACSSM Tech. Memo. 81, pp. 244–253.

BERGLUND, G. and B. AKELYON. 1949. Drainage by explosives. Lantmannen, No. 34, Bord na Mona E. S. Transl. No. 203 (1951), Dublin.

BRAWNER, C. O. and R. O. DARBY. 1962. Culvert selection in muskeg areas. Proc. Eighth Muskeg Res. Conf., NRC, ACSSM Tech. Memo. 74, pp. 84–99.

BROWN, R. J. E. 1965. Distribution of permafrost in the discontinuous zone of Western Canada. Proc. Can. Reg. Permafrost Conf., NRC, ACSSM Tech. Memo. 86, pp. 1–14.

BROWN, R. J. E. and G. H. JOHNSTON. 1964. Permafrost and related engineering problems. Endeavour, Vol. XXIII, No. 89, pp. 66–72.

CASAGRANDE, L. 1966. Construction of embankments across peaty soils. J. Boston Soc. Civil Engrs., Vol. 53, No. 3, pp. 272–317.

CUTHBERTSON, J. A. 1959. Aspects of ditching and drainage techniques in muskeg areas. Proc. Fifth Muskeg Res. Conf., NRC, ACSSM Tech. Memo. 71, pp. 62–67.

ENGINEERING NEWS-RECORD. 1942. Building airports on New England's bogs. Vol. 129, No. 5, pp. 49–51.

FALKLAND ISLANDS AGRICULTURAL DEPARTMENT. 1967. Grassland improvement. Proc. Conference held at Stanley (July 1967). 30 pp.

FENCO. 1960. Building on peat in Burnaby: a summary engineering report. Foundation of Canada Eng. Corp. Ltd., Vancouver, BC, 6 pp.

GOODMAN, L. J. and C. N. LEE. 1962. Laboratory and field data on engineering characteristics of some peat soils. Proc. Eighth Muskeg Res. Conf., NRC, ACSSM Tech. Memo. 74, pp. 107–129.

GUSTAFSON, C. J. 1965. Installing utilities in muskeg. Proc. Reg. Seminar on Organic Terrain Problems: Appendix B, in Proc. Tenth Muskeg Res. Conf., NRC, ACSSM Tech. Memo. 85, pp. B9–B12.

HARDY, R. M. 1967. Unpublished data.

HEALY, J. V. 1961. Organic terrain reclamation in Newfoundland. Proc. Seventh Muskeg Res. Conf., NRC, ACSSM Tech. Memo. 71, pp. 78–92.

HEIKURAINEN, L. 1963. Results of draining peat lands for forestry in Finland. Paper presented at Second Intern. Peat Congress, Leningrad, USSR (manuscript).

HEMSTOCK, R. A. 1967. Personal communication.

HIGHWAYS AND BRIDGES. 1942. Constructing the Whatcom County, Washington, airport. Vol. 8, No. 410, p. 4.

HUNTER, C. F. 1967. Consolidation of recent alluvial deposits in British Columbia. Unpubl. M.SC. thesis, University of Alberta. 93 leaves.

JOHNSTON, R. N. and G. A. HILLS. 1956. The need for rehabilitation of organic terrain in Ontario with special reference to reforestation. Proc. Eastern Muskeg Res. Meeting, NRC, ACSSM Tech. Memo. 42, pp. 46–51.

KAPP, M. S., D. L. YORK, A. ARONOWITZ, and H. SITOMER. 1966. Construction on marshland deposits: treatment and results. Highway Res. Board, Highway Res. Rec. No. 133, Washington, DC, pp. 1–22.

LAKE, J. R. 1961. Investigations of the problem of constructing roads on peat in Scotland. Proc. Seventh Muskeg Res. Conf., NRC, ACSSM Tech Memo. 71, pp. 133–148.

LEA, F. M. 1956.   The chemistry of cement and concrete (rev. ed. of Lea and Desch), Edward Arnold (Publishers) Ltd., London. 637 pp.

MACFARLANE, I. C. 1965. The corrosiveness of muskeg waters: a review. Can. Geotech. J., Vol. 2, No. 4, pp. 327–336.

MAINLAND, G. 1962. Corrosion in muskeg. Proc. Eighth Muskeg Res. Conf., NRC, ACSSM Tech. Memo. 74, pp. 100–106.

MARKOWSKY, M. and N. J. MCMURTRIE. 1961. Transmission line structure foundations in weak soils. Proc. Sixth Muskeg Res. Conf., NRC, ACSSM Tech. Memo. 67, pp. 32–37.

MARTIN, J. 1965. Building in muskeg. Proc. Reg. Seminar on Organic Terrain Problems: Appendix B, in Proc. Tenth Muskeg Res. Conf., NRC, ACSSM Tech. Memo. 85, pp. B5–B8.

MCLURE, G. E. 1967. Personal communication.

MOLLARD, J. D. 1961. Guides for the interpretation of muskeg and permafrost conditions from aerial photographs. Proc. Sixth Muskeg Res. Conf., NRC, ACSSM Tech. Memo. 67, pp. 67–87.

MOLLARD, J. D. and J. A. PIHLAINEN. 1963. Air photo interpretation applied to road selection in the Arctic. Proc. Permafrost Intern. Conf., Lafayette, Indiana. Nat. Acad. Sci.–NRC Publ. No. 1287, Washington, DC, pp. 20–25.

MULLER, S. W. 1947. Permafrost or permanently frozen ground and related engineering problems. J. W. Edwards, Inc., Ann Arbor, Michigan, 211 pp.

MUNOZ, C. C. 1948. Heavy tractors compact swamp fill for airport enlargement. Engineering News-Record, Vol. 140, No. 24, pp. 948–949.

PIHLAINEN, J. A. 1963. A review of muskeg and its associated engineering problems. US Army Materiel Command, Cold Regions Res. and Eng. Lab., Tech. Rept. 97, Hanover, NH, 56 pp.

POWERS, T. C., L. E. COPELAND, and H. M. MANN. 1959. Capillary continuity or discontinuity of cement paste. Portland Cement Assoc., Res. and Development Lab., Bull. 110, 11 pp.

P'YAVCHENKO, N. I. 1963. The drying of bogs as a means of transforming nature and developing the forest resources of the Siberian North. Problemy Severa, No. 7, pp. 55–64. Problems of the North (Transl. by NRC of Canada), 1964, No. 7, pp. 57–66.

ROADS AND STREETS. 1944. Runway extensions a mucking job. Vol. 87, No. 5, pp. 55–60.

SAVAGE, J. E. 1965. Location and construction of roads in the discontinuous permafrost zone, Mackenzie District, Northwest Territories, Proc. Can. Reg. Permafrost Conf., NRC, ACSSM Tech. Memo. 86, pp. 119–131.

STANWOOD, L. 1958. Moving a million yards of muskeg. Construction World, Vol. 14, No. 2, pp. 30–33.

THOMSON, S. 1963. Icings on the Alaska Highway. Proc. Permafrost Intern. Conf., Lafayette, Indiana. Nat. Acad. Sci.–NRC Publ. No. 1287, Washington, DC, pp. 526–529.

WAHLS, H. E. 1962. Analysis of primary and secondary consolidation. Proc. Am. Soc. Civil Engrs., J. Soil Mech. Foundations Div., Vol. 88, No. SM6, Pt. 1, pp. 207–231.

8 Trafficability and Vehicle Mobility

Chapter Co-ordinator*
R. A. HEMSTOCK

8.1 INTRODUCTION

Transportation is an essential factor in the commerce of all countries. Large countries such as Canada with widely distributed natural resources are especially dependent on effective transportation. For economic viability the transportation system must be able to meet the needs of the extractive industries, regardless of the distances and terrains involved.

In Canada, over the years, a transportation system has been slowly emerging in which a certain general pattern has been repeated a number of times. This pattern begins with man exploring an area on foot, on horseback, or in a small boat. With the finding of resources, be they farm lands, forest, or minerals, comes a need for mass transportation. Obviously, the more permanent transportation structures such as railways and highways must wait until the value of these resources has been established. Therefore, mass transportation in its early forms has traditionally been cross-country, as exemplified by the canoes of the fur traders, the river-driven logging of eastern Canada, the Red River carts of the prairies followed by the horse-drawn wagons and sleighs. The last two mentioned carried goods and people thousands of miles and were essential in the development of the Canadian prairies. Today, tractor trains, off-road tracked and wheeled vehicles, helicopters, and STOL aircraft are filling a similar role in the north's newly developing areas.

Understandably, the first resources to be developed have been those in places where the surface terrains readily permit the passage of transportation equipment. Sometimes, indeed, richer resources have had to be bypassed because there was no suitable equipment to carry the required machinery into the site and the product out to the market.

Among the numerous difficult terrains that have delayed exploitation of some of Canada's resources, organic terrain, commonly referred to as "muskeg," holds a special place. Except in winter, it has long been an insurmountable barrier to transportation equipment and, to a slightly lesser extent, transportation

*Chapter Consultants: D. Campbell, R. Newcombe, J. E. Rymes, J. G. Thomson.

structures. Recently, very modest research programmes along with some intuitive vehicle-design projects have provided the basis for a varied group of vehicles which can more or less cope with the transportation requirements over most organic terrain. Further quantitative data on muskeg physiographic and mechanical characteristics can broaden the basis for improved vehicle design and hence better transportation systems.

8.2   TERRAIN CHARACTERISTICS

Certain features of a terrain are bound to affect the ability of any type of vehicle to traverse it. A few of the more obvious are soil characteristics such as bearing capacity, shear strength, depth and water content; topography such as forests, hills, snowdrifts, ice ridges, lakes and rivers; and seasonal variations such as frozen or flooded ground. It is obvious that some of these, depending on their severity, may immobilize certain vehicles altogether, while others may only slow them down.

Certain features are "island barriers" – that is, they cannot be crossed, but they may be circumnavigated. Lakes act as island barriers to all but amphibious vehicles, and at first muskeg did also, with men, railways, highways, and off-road vehicles all going around it.

Other barrier types are chain or boundary barriers such as rivers or, on a larger scale, mountain ranges. They do not necessarily influence travel in terms of an areal factor, but may effectively limit access by creating a boundary. Alternatively, they may demand performance capabilities in a vehicle design which would not be required for its major role.

Blanket barriers such as snow cover or continuous muskeg influence travel areally. This type of barrier may prohibit access to areas beyond as well as within its own boundaries.

Finally, the seasons themselves can affect cross-country travel. What was a barrier in summer may become a highway when frozen in winter, as in the case of lakes, some rivers, and some muskegs. On the other hand, certain terrains that are passable in summer become impassable in winter because of deep snow, ice ridges, or, with some types of muskeg, severe surface hummocking.

8.3   TERRAIN TRAFFICABILITY

Terrain trafficability may be defined as the extent to which a terrain permits the passage of vehicles. From this definition it follows that a terrain has a poor trafficability characteristic with respect to wheeled or tracked vehicles if it provides little resistance to penetration by the vehicle (i.e., low bearing strength resulting in vehicle sinkage into the soil), or if it has low capacity for reacting to the tractive thrust of the vehicle (i.e., low internal sheer strength resulting in low vehicle tractive effort). A terrain may also have "geometric obstacles" such as trees, hills, and hummocks. Organic terrain often exhibits all of these characteristics at once.

Soil trafficability deals only with the portion of the terrain at or below the ground surface, and for many terrains it is all that need be considered. In the case of organic terrain it is often useful to consider the soil portion separately, though many other factors are also important.

8.4   CLASSIFICATION

Until organic terrain could be classified, there was no real basis on which to conduct studies of vehicle performance in muskeg. Some of the earliest vehicle proposals clearly reflect a lack of appreciation of the range of variations in muskeg; in other words, their operating scope was very limited.

As recently as 1952, in the absence of a classification system, there was still a kind of mythology surrounding the problem of muskeg trafficability. With no factual data on muskeg, only the worst features were remembered, and even these memories often reflected inaccurate observation. The Radforth (1952) classification, one of the early results of muskeg research, provides a base on which physical measurements can be made and correlated. It includes nine major muskeg cover types and an indefinite number of combinations of these (see Chapter 2). From this starting point one can begin to develop an understanding of the terrain factors affecting vehicles in muskeg.

Although a particular organic terrain may be classified as unable to support vehicular traffic, it will often be found that the limiting factor is not the organic material. Radforth (1958) has pointed out that the physical condition of organic terrain "is governed by the structure of the peat it contains, and its related mineral sub-layer, considered in relation to the features and the surface vegetation with which the peat co-exists." In assessing the trafficability of muskeg terrain all four factors are important.

In any muskeg the layer of live vegetation provides much of its "mat strength." The living, flexible roots, stems, and branches tend to interlock, thus spreading and supporting the vehicle load through their own tensile strength. At the same time the size of plants in the living layer (for example, trees) may impede the progress of certain vehicles.

The structure of the peat layer, which occurs immediately below the living layer, is described in Chapter 2. Both structure and water content affect the peat's shear strength and bearing capacity. These characteristics, in turn, can be correlated with trafficability of the terrain.

The mineral layer immediately underlying the peat may be a critical trafficability factor, particularly when the peat layer is relatively shallow. In this case vehicles often bog down as a result of slippage along the interface of the peat and mineral soil. Alternatively, a shear or bearing failure may occur in the mineral soil, partly because of the high water content supplied from the organic layer.

8.5 VEHICLE MOBILITY

Mobility has been defined as "the quality or state of being mobile" or as having "capacity or facility for movement." According to these definitions, the ability of a vehicle to move in an environment may be equated to vehicle mobility. The mere ability to move, however, is not usually of much interest, since it does not provide a measure of how much useful work can be done or at what rate.

8.6 DRAWBAR-PULL TEST

Early quantitative studies of vehicle mobility were carried out to produce data pertinent to the vehicle's intended use. These studies began when off-road machines were mainly tractors of one kind or another. Since the basic purpose was to tow or push a load, the drawbar-pull test was devised. That test was a direct measure of the capability of the vehicle to perform its design function under the test soil conditions. Because many of these tractors were for farm or logging use, the range of soil conditions to be considered was not large. Certain "standard" soils, however, bearing regional names, demanded test media when it was found that the results of tests in one area did not necessarily represent the performance in another. Over the years, as the soil characteristics that contribute to vehicle performance – such as depth, density, grain size, and water content – were identified, it became possible to run "controlled" tests. Performance in other soils could then be predicted on the basis of the test results from the "control" soils.

The advent of self-propelled earth-moving machinery did little to change the test method, since the requirement was still for drawbar pull. The range of "standard" soils, however, had to be extended.

Military vehicles (other than artillery tractors) were an early departure from the purely drawbar-pull type. Instead, they had to be able to accelerate to much higher speeds, climb hills, and manoeuvre in a wide range of adverse soil conditions. The drawbar-pull test was used to evaluate them, on the basis that any traction more than was needed to move the vehicle on a horizontal surface at uniform speed could be diverted to these other purposes.

8.7 MEASURING VEHICLE CAPABILITY

With military and civilian tracked and wheeled vehicles for use in off-road transportation, an additional factor has to be considered. This is the rate at which goods can be moved. For many years this factor has been expressed in two ways: as ton-miles per hour while travelling in the design media, and as hours required to go from point A to point B without regard to the route used. Clearly, these are not alternative definitions of one capability. With the second criterion there is obviously the risk of subjectivity in the choice of test location, and this method has not gained wide support.

Recently, attempts have been made to devise a more meaningful test of mobility than the drawbar-pull but none of these provides the full story. Neither, however, does the drawbar-pull test, as the following example will illustrate.

In preliminary tests Liston and Hanamoto (1966) found that the drawbar-pull/weight ratio was not related to vehicle performance. In other words, soft-soil performance and vehicle mobility are not equivalent. Their first tests were made by timing five different vehicles negotiating a test course laid out so as to duplicate the manoeuvre requirements of off-road terrain.

In these tests, a vehicle with only average soft-soil performance required significantly less time to traverse the course and the authors conclude: "It is obvious that an extremely low, zero, or negative drawbar-pull/weight ratio would result in poor performance. However, it is also apparent that under the conditions tested, the drawbar-pull/weight ratio is primarily a go, no-go indicator. This observation is a departure from accepted thinking – drawbar-pull/weight ratio has been accredited as an indicator of both go and no-go and how much 'go' because it was thought to indicate the force available for hill climbing, load pulling, or accelerating" (Liston and Hanamoto 1966).

The quotation is admittedly out of context. The report does not set any particular limits, however, on the applicability of the contents. The valid conclusion to be drawn from the work of these authors is that the drawbar-pull test does not allow for geometric obstacles. Without excess traction, or drawbar-pull capability, the test vehicles would never have arrived at the test site or moved through the test area. As was noted in an earlier case, this proposed test method is prone to subjectivity in laying out the test course. Results obtained from the procedure, therefore, should not be considered a sufficient basis for vehicle mobility ratings.

8.8 VEHICLE PERFORMANCE IN MUSKEG

From the foregoing, it is apparent that a great deal of effort has gone toward the development and use of techniques for evaluating vehicle mobility quantitatively and qualitatively. Originally, the work was restricted to mineral soils; later it was shown that some of the same techniques could produce similar data in deep snow. It remained to demonstrate that the methods could also be used in muskeg.

Thomson (1961) first described the results of controlled experiments with vehicles in muskeg. He found that: (1) standard test procedures for testing vehicles in other soils and in snow were applicable in muskeg; (2) the test results indicated a direct relationship between muskeg mechanics and soil mechanics.

It was shown in these tests that vehicle performance is sensitive to the conditions of the surface layers of muskeg at the time of the test. Moreover, in many FI muskegs track-contact pressures of up to 5 pounds per square inch (0.35 kg per sq cm) might be satisfactory for well-designed and carefully loaded vehicles. This latter finding came as a distinct surprise to many operators, who had thought that the whole secret of transportation across muskeg was to have the lowest possible average ground pressure.

Thomson (1961) was able to correlate the shear strengths of peat, obtained by shear vanes, with vehicle-test results. Lanes of high shear strength produced the best vehicle performance. Moreover, the shear-strength data could be correlated with the entire vehicle "pull slip" curve. It was also shown that muskeg of a specific classification did not have an exclusive shear strength. Water content and peat depth appeared to influence shear strength and hence vehicle performance.

Thomson concluded from the data obtained:

1. . . . the performance of a vehicle on or in muskeg is qualitatively the same as that of a vehicle in any other soft or loose media, be it clay, sand, or snow. It must therefore be concluded that the general principles of vehicle design already established for other media and of the science of soil mechanics are directly applicable to muskeg;

2. At least a preliminary correlation between shear strength as measured with the shear vane and net vehicle traction has been established. The correlation depends on two assumptions;

3. As long as sufficient bearing capacity is available to prevent ultimate sinkage of the vehicle the net traction of the vehicle can probably be predicted from the shear strength of the muskeg surface mat;

4. A second correlation between the shear vane and vehicle performance has been established through the use of three lanes of varying shear strength which produced consistent net traction variations within the accuracy of the test method;

5. The shear vane as used is not capable of developing sufficient information on the shear strengths of muskegs to permit absolute correlation with vehicle test results. Continuous readings of the stress–strain curve is required from zero strain until a strain is reached beyond which there is no further decrease of stress;

6. The shear stress values reported are probably strain rate dependent. It has been shown in previous work (Thomson 1957) that the net tractive performance of a vehicle is also strain rate dependent. To obtain the best possible correlation it may be necessary to make repetitive readings of the shear strength at various strain rates. These values would then be applied to the strain rate beneath the vehicle tracks at various absolute track speeds, each to each;

7. The removal of one-half of the track grousers significantly improves the mobility of the test vehicle;

8. Reducing the number of grousers must result in better utilization of the available shear strength of the muskeg through a better interlock between the grouser bars and the muskeg mat fibres;

9. Types FI, F and I muskegs do not present an insurmountable barrier to vehicle travel. This conclusion is further strengthened by operation of the same vehicle in another FI muskeg which was much deeper, had a less well developed surface mat,

and had much lower shear strength at depth than that used for the instrumented vehicle tests;

10. Sufficient information has been developed to show that the performance of future vehicles can be compared with that from other vehicles measured at another time and place. The shear vane data will provide the link;

11. The test vehicle, in the original form, is underpowered for operation in types FI, F and I muskeg.

It should be emphasized that this work was concerned essentially with the mobility of a particular tracked vehicle. Once it has been established that a vehicle has enough mobility to do a job, then other factors must be considered to determine the efficiency or suitability of that vehicle as compared with other vehicles of comparable mobility.

8.9   VEHICLES AND THEIR ENVIRONMENT

A broad approach aimed at a better understanding of the relation between a vehicle and its physical environment is being taken at the Land Locomotion Laboratory in Warren, Michigan (cf. Bogdanoff *et al.* 1965–66). The major problem is recognized as being associated with soft soils, but it is also stated (Reece 1964) that the study is in its infancy and cannot deal at present with even a wide range of deep uniform soils.

Reece (1964) lists five approaches to the problem as follows: "(a) a theoretical approach based on civil engineering soil mechanics; (b) an analytical approach using semi-empirical soil stress–deformation relationships; (c) index systems based on attempting to describe soil characteristics by means of a single simple measuring device; (d) model experiments using dimensional analysis to systematize the results; (e) a rigorous mathematical approach based on the theory of plasticity." (*Ibid.* 1964.)

The soil mechanics approach was begun in Britain when vehicles were becoming immobilized during World War II. Bekker (1956) followed this approach in his early work, but later broadened it to include soil-stress/deformation relationships. A cone penetrometer (see Section 5.2(2)) has been used successfully by the US Army Engineers Waterways Experiment Station at Vicksburg, Mississippi, to predict trafficability in the field for soft soils and snow. In 1959 they reported that they had also been able to apply the cone penetrometer to predicting trafficability/vehicle mobility in muskeg (US Army 1959). This penetrometer method is now used quite extensively for predicting vehicle performance.

Model work has been hampered by the lack of suitable soil-describing constants. This technique is not likely to yield results until a firm theoretical foundation is available.

The mathematical approach based on the theory of plasticity has not worked because soils do not behave as ideal plastic materials. Reece (1964) believes that this approach might be fruitful except that "applied mathematicians . . . are primarily concerned with the construction of logically correct systems of ever-increasing complexity rather than with the solution of engineering problems."

The preceding paragraphs indicate that quantitative data on vehicle perform-ance in muskeg are attainable and that new techniques for determining mobility characteristics are being studied. It is not possible to say whether these new tech-niques will be applicable to vehicle mobility in muskeg; the drawbar-pull test may still have to suffice for many years. Whichever approach proves more fruitful, it must be noted that both are concerned with the soil rather than with the terrain.

With reference to the terrain, therefore, a few other mobility factors must be considered. In organic terrain, topography is a factor that affects mobility. Both large-scale features, such as steeply sloping muskeg, and small-scale features, such as hummocks, ice mounds, pot-holes, and pond margins, must be considered. Some vehicles may be immobilized by such geometric obstacles and even the more sophisticated may be slowed down by rough terrain.

Terrain roughness strongly influences the selection of a vehicle and the final costs of any off-road transportation project. Unlike inadequate traction, rough-ness is not normally a go – no-go factor. Rather, it is a speed-limiting factor that concerns the interaction of vehicle length and suspension characteristics with the pitch and amplitude of the terrain surface variations. Relatively little attention has been paid to how roughness affects the success or failure of off-road transporta-tion systems. Techniques to measure the surface roughness of muskegs and relate it to the vehicle have not been developed, although the types of muskeg in which it commonly occurs and its mechanical causes were described by Radforth several years ago (1952, 1954).

The US Army Tank-Automotive Center in Warren, Michigan, is investigating ground roughness by statistical methods (Bogdanoff *et al.* 1966; Kozin *et al.* 1963). It has found that two-dimensional ground spectra can be obtained which give all the information needed to predict roughness both for the tracks of a vehicle and for travel in either direction. This is part of the basic information required to develop a theory of land locomotion.

In Canada, ground-roughness research has been started by J. R. Radforth (1967). The emphasis is on measurement techniques and on converting data to a form suitable for use with a computer.

It is hoped that research into terrain roughness will lead to more appropriate designs for tracked and wheeled vehicles. The aim is to find methods that will allow ground roughness to be estimated from aerial photographs, muskeg classifi-cation maps, or topographical maps.

8.10 VEHICLE DESIGN CONSIDERATIONS

Vehicles should be designed to do the best possible job in the terrain where they have to operate, within the limits of design principles. It must be emphasized that any practical vehicle must not only be effective in muskeg, it must also be able to do useful work in a variety of terrains in all seasons. Generally, when a vehicle is designed for a specific type of terrain, it tends to become either limited or exotic and is not likely to perform well in other terrains. For example, most amphibious vehicles either do not swim very well, or if they do, behave literally "like a duck out of water" when on land.

In view of the diversity of terrain, it seems logical to assume that no one vehicle will ever be developed which will be equally or even acceptably efficient everywhere – at least not within realistic economic constraints. The most successful vehicle in any one area, however, will be that which best combines the requirements imposed by the various types of terrain encountered. It need not, then, be the one with the greatest mobility, but it will be the one that offers the best combination of mobility, speed, ride characteristics, serviceability, reliability, safety, load-carrying ability, and cost of service. The study of these interrelations is known as performance systems analysis. There is a great deal of research that can be done in this area, which it is hoped will lead to increasingly better vehicles.

In Canada, the jobs to be done by cross-country vehicles are generally related to natural-resources industries. These include timber harvesting, power line or pipeline installation and maintenance, freight hauls for the mining and petroleum industries, and exploration. In some cases loads are light – perhaps only personnel – in others large tonnages or very heavy single pieces will be involved. Obviously, just as no one vehicle can be equally proficient in all terrains, no one vehicle or even family of vehicles can be expected to do the wide range of jobs encompassed by off-road transportation systems in Canada.

8.11 PRACTICAL VEHICLE CHARACTERISTICS

The characteristics required for off-road vehicles are embodied in such words as mobility, speed, ruggedness, maintenance costs, useful life, comfort, safety, serviceability, and total cost of transportation. For a specific vehicle, these qualities may sometimes be inferred by a study of design features such as: ground pressure; power-to-weight ratio; tracks or wheels; length-to-width ratio of tracks and/or vehicle; steering mechanism; approach angle of the front of the track or wheel size; suspension system; front or rear drive; flotation (amphibious) capabilities; maximum and typical operating speeds; transmission, gear ratios, and type; driver comfort and safety features.

Research programmes have shown the following to be important vehicle-design considerations:

1. The longitudinal location of the dynamic centre of gravity affects the vehicle contact pressures. The nominal unit ground pressure does not indicate the actual track-contact pressures except where their dynamic centre of gravity acts through the centre of the track contact. It can be shown that some vehicles with an average ground pressure of 2 pounds per square inch (0.14 kg per sq cm), based on the

gross vehicle weight and the full track-contact length, actually operate at an average of 7 pounds per square inch (0.49 kg per sq cm) on the aft foot or two (0.30 or 0.61 m) of the track. Redistributing the weight can improve the tractive performance by more than 50 per cent.

2. Long vehicles can operate successfully at higher contact pressures than short units. This is a result of the interaction of vehicle and soil geometry. In practice this principle must be applied by using units en train. For best results, care must be taken to ensure that the units act as a co-ordinated whole rather than as coupled individuals. Long narrow track also reduces the bulldozing effect, as compared to a short wide track of the same area.

3. Skid-steering systems such as the braked, clutch-brake, and controlled differentials, the military cross-drive, and the various forms of torque-proportioning final drives, apply an extra mobility demand on the soil. Steering systems that are independent of the shear strength of the soil, such as the centre-frame joint-articulated system, are more positive and add to the vehicle's overall mobility, since the steering force does not require a strong reaction from the soil.

4. Minimum belly width and maximum practical belly height reduce drag at deep penetration and therefore increase net traction.

5. Track length is more effective than track width in increasing net traction.

The foregoing are some of the factors that influence the vehicle mobility. To make it practical and convenient to operate, the design must include other factors that can be deduced from observation of full-scale vehicles in action. The following examples illustrate this type of design variable:

6. A positive, accurate, low-effort steering system is essential. The vehicle response must be proportional to the motion of the steering-wheel or lever rather than to the elapsed time since steering was initiated. The turning rate must be fast enough to manoeuvre the vehicle at full speed.

7. A rugged but deep suspension, compatible with the maximum vehicle speed and expected trail roughness, is essential. A rough-riding suspension will lengthen trip time and increase repair costs. This is a complex problem involving relations between deflection, jounce and rebound, control frequency of the suspension, and the system of one bearing wheel with respect to another. Vehicle speed must be coupled with control ability.

8. The overall design must be rugged enough for, and dimensionally suited to, the application. Although the vehicle tare weight should be as low as possible, it must be high enough to provide adequate strength.

9. The vehicle must be adequately powered. The power plant must meet two conditions: it must have enough power to maintain top speed in fairly easy going, and enough torque to maintain headway under all the soil and hill conditions to be encountered in the terrain.

10. "Hot shift" transmissions are essential. A slow gear-change while crossing muskeg can cause the vehicle to bog down. An unnecessarily low gear wastes time. Torque converters are always convenient but waste engine power when operating

at the high torques and wide throttle openings common to off-road vehicles. On the other hand, this drawback may be more than offset by the advantage gained through a smooth torque increase as soil conditions deteriorate.

11. The entire driveline must be able to accept the peak engine torque continuously. This is opposed to truck practice, where the maximum driveline loadings are determined by the weight on the powered wheels and the coefficient of friction of the tire on the soil or pavement.

Before deciding which of the desirable characteristics should be incorporated in a proposed vehicle, some thought must be given to the economic environment. Generally, the economic factors to be considered are: first cost, operating cost, availability rate, utilization rate, life requirement, and loss of income resulting from down-time. Trade-offs must usually be made. For example, a vehicle required for intermittent use during a short exploration period might be satisfactory if it could receive the maintenance it needed during the expected standby periods. In that case, a low-first-cost, high-maintenance-cost vehicle might be satisfactory. On the other hand, a requirement for high utilization in a project that must be shut down if the transportation system fails calls for a high-first-cost, low-maintenance-cost or premium-quality vehicle.

8.12 EXISTING VEHICLES

The manufacture of muskeg vehicles is a relatively new Canadian industry, in which most of the progress has been made since the late nineteen fifties. The vehicles have been widely accepted in Canada, and are now used from British Columbia to Newfoundland and from the prairies to the Arctic islands. Oil exploration by seismic methods, for example, had been impossible in much of the north on a year-round basis until the advent of these vehicles. Now it is accepted as almost routine.

Canadian vehicles have also enjoyed wide acceptance in many other countries for cross-country work. They have been used in such highly varying terrains as snow in Antarctica, desert sand in Australia, tropical jungles in Venezuela, and the interior flood plains (Llanos) of Colombia. This wide range of applications emphasizes the similarity of many problems that show up in cross-country work over terrains that are quite diverse. In terms of the ability to support traffic, of course, all soft terrains have similar characteristics, although the effect on the vehicle varies.

Tracked vehicles enjoying acceptance in the market today have been designed usually for specific requirements in either the forestry or the petroleum industry. Most feature continuously flexible band tracks. Enough track area is usually fitted to produce low average ground pressures of from 1.5 to 3.5 pounds per square inch (0.10 to 0.25 kg per sq cm). Both front and rear drive are used. Some vehicles feature an approach angle, while others do not.

Vehicles with a carrying capability of up to 8 tons (8.82 tonnes) may use some form of skid steering. Machines in the 4–20 ton (4.41–22.1 tonnes) capacity range may use either centre-frame joint-articulated or wagon steering. Amphibious characteristics have not been in demand, possibly because of the compromise required in soft-ground performance.

Gasoline or diesel power plants are available, and the choice here is often based on the user's preference – or the need for fuel compatible with that used by other vehicles in the fleet or area. The choice of conventional, automatic, or power-shift transmission is again often based on user preference. Many vehicles suffer from underpowering and relatively poor matching of engine and transmission to the vehicle needs.

Maintenance costs on all track vehicles are relatively high. This is due partly to the rough conditions under which they operate and partly to the complexity and cost of the track systems. Nevertheless, since the early sixties there has been a marked improvement. Manufacturers realize that high maintenance costs are now the main item in discouraging wider use of their vehicles, and consequently are improving their products.

8.13   VEHICLE MAINTENANCE

As has been pointed out already, many factors affect the performance and therefore the maintenance requirements of the muskeg vehicle. The very nature of the muskeg type of terrain, for which the vehicle is best suited, does not lend itself to maintenance practices where and as required, but does predicate that a preventive-type maintenance be planned.

By way of illustration, a simple wheel-bearing failure can result in a very wet and unpleasant repair procedure if circumstances dictate that repairs be made "mid-swamp." A track failure while descending or ascending a steep grade can dictate a difficult repair procedure even if the operator is able to control the machine with the one remaining track. A simple track grouser break can puncture a tire, and under some conditions puncture several tires, before the operator can safely stop the machine. Very often the punctured tire results in a ruined tire rim. Although repairs can be made under the above circumstances, it must be realized that, in addition to the inconvenience and unscheduled out-of-service time, the cost in material to make repairs "as and where needed" is very much higher than for scheduled repairs.

These illustrations exemplify the high costs and unpleasant aspects of field repairs, and these aspects must be given their fair values in relation to vehicle maintenance costs resulting from preventive maintenance practices.

(1)   *Scheduling of Maintenance*
Scheduling of maintenance functions can best be originated from the individual

manufacturer's recommendations for the particular operating conditions. Since the manufacturer is not a continuing observer of a particular operation, the scheduling of maintenance functions must frequently be adjusted to yield the availability and reliability that the job demands. It must also be recognized that, if an increase in availability and reliability is required, an increase in costs of maintenance can be expected. At some point, peculiar to each machine and to its operating conditions, increasing the scheduling of maintenance functions results in sharply diminished returns for the extra costs expended.

Alternatively, reducing the scheduling of maintenance functions may result in a better return of the costs expended provided that the lesser available-for-work time and reliability can be tolerated. Somewhere between these two extremes lies an optimum scheduling of maintenance for each particular job and its operating conditions. This optimum scheduling can be determined and maintained at the optimum level for any particular job through an occasional trial at a level above or below the accepted optimum level to "proof test" the accuracy of the practice.

(2) Inspections to Schedule Repairs
The construction of the muskeg vehicle and its type of service makes prediction of the service life left in any of the components of the vehicle difficult to forecast for any appreciable period of time. Inspections of the vehicle components must be performed at frequent intervals, therefore, by a knowledgeable person so that maximum availability of the vehicle can be obtained.

An inspector's judgment which gives a reasonably accurate prediction of the life remaining in the vehicle components results only from his diligent practice of inspections, predictions, and observation of the results obtained. Development of this judgment in any individual, therefore, is a result of time and exposure to the vehicle and its operating conditions.

(3) Inspections and Lubrication
Inspection and lubrication of the vehicle can often be best performed at the same time. The performance of the lubrication and these particular items of inspection service dictate that the machine's components be under the eye of the inspector in an orderly manner and on a regular basis.

Some items of inspection and lubrication must be considered as inseparable. By way of illustration, the flushed lubricant from a particular wheel bearing may show unusual contamination. This observation is a part of an inspection of the vehicle and should result in a plan for an effective repair at the earliest possible time. If immediate repair is not possible, the lubrication service interval for that wheel bearing should be reduced to effect a level of lubricant contamination which will prolong the life of the bearing until repairs can be made.

Although lubrication and inspection must be considered as inseparable, many components of the vehicle which do not require lubrication must be inspected and some lubricated components may require inspection more frequently than they require lubrication. Belt-type tracks, grousers, tires, and suspensions are items that must be closely observed at frequent intervals to determine whether they are wearing at unusual rates or showing signs of damage from certain operating conditions.

(4) *Lubrication*

The lubricants represent the smallest item in the operating costs, but their effective use will have the greatest single effect in reducing the costs of maintenance and maintaining a high available-for-work factor. The nature of the wet peat, soils, and sands in which many components of the vehicle are submerged for long periods of time and exposed to on an almost continuing basis, cannot be sealed out with economic present-day sealing mechanisms unless supported by frequent lubrication. Those components of the vehicle which are lubricated with liquid lubricants (motor transmissions, gear reduction units, and differentials) must be regularly and frequently checked for contamination of the lubricants. Generally, this is best accomplished by sampling at the lowest area of the sump (the drain is normally arranged to permit this) after the vehicle has been stopped for an hour or more to allow settling of the contaminants and moisture. The gear oils and motor oils must be watched carefully and frequently for changes in colour or other characteristics which may indicate contamination not indicated by a sump check. Generally, the lubricant changes should be regulated by the manufacturer's recommended change interval, unless inspections show that operating conditions are resulting in lubricant conditions that require more frequent changes.

Those components of the vehicle normally provided with vents or breathers that can be plugged for special conditions must be watched with care when they are being operated without vents. The lubricants and the air in the components expand and contract with temperature change and when breathers are plugged, oil seals can be ruptured.

The grease lubricated components (namely, all the support and lubricated track components, control linkage, etc.) depend on the grease not only as a lubricant but also as a medium to flush out contaminants and to provide a backup for the oil seal members. Care in the selection of a grease lubricant that provides good resistance to water pumpability at low temperatures and is stable within a wide range of operating temperatures will be well repaid.

(5) *Maintenance Records*

The records of the maintenance work performed on a vehicle are of invaluable assistance in the knowledgeable adjustment of maintenance practice to the demonstrated requirements of the muskeg vehicle. Records need not be detailed or difficult to maintain in order to provide the facts needed at any future time to re-adjust maintenance to correct its shortcomings or even to reduce its frequency to suit the equipment's demonstrated abilities.

In summary, the nature of the muskeg vehicle and its work demand maintenance practices of a higher order than most other vehicles. The frequency of maintenance practices and inspections must be determined from usage on each particular job situation and adjusted to suit the needs demonstrated by the vehicle.

The useful and reliable life remaining in components of a particular vehicle, in a particular work application, must be determined from usage history and the results of frequent inspections. Occasionally the life expectancy of various components should be proved by a trial extension of the current practice.

Records are necessary to provide guidance in the adjustment of maintenance. Inasmuch as the muskeg-type vehicle is a sole economic means of transport over many types of muskeg and soft ground, adequate maintenance is the only means of keeping it economical and dependable.

8.14 SOME VEHICLES AVAILABLE IN CANADA

Figures 8.1 to 8.10 illustrate some of the tracked vehicles which have been developed in Canada. No attempt has been made to list all the vehicles available; rather the purpose is to show the range and characteristics of several vehicles. *This should not in any way be construed as an endorsation of any particular vehicle.* To lend perspective to the design principles, pertinent specifications are given for the vehicles illustrated.

FIGURE 8.1   Bombardier model S Muskeg Carrier

Length: 140 in.
Width: 87 in.
Weight: 6000 lb
Payload: 6000 lb
Track pressure loaded: 2 psi
Speed (max): 16 mph
Engine: 125 hp @ 3600 rpm
Track width: 28 in.
Steering: controlled differential

FIGURE 8.2   Flex-track model 575 tracked carrier

Length: 190 in.
Width: 97 in.
Weight: 8500 lb
Fording depth: 50 in.
Payload: 3500 lb
Ground pressure, loaded: 1.6 psi
Speed (max): 15 mph
Engine: 6 cylinder, 225 cu in.
Steering: controlled differential

FIGURE 8.3   Flex-track model 800 tracked carrier

Length: 265 in.
Width with tracks: 117 in.
Gross weight: 31,000 lb
Fording depth: 54 in.
Payload: 13,000 lb
Ground pressure, loaded: 2.47 psi
Speed: 15 mph
Engine: 8 cylinder, 361 cu in.
Steering: controlled differential

FIGURE 8.4   Foremost tracked carrier model 6T

Load: 12,000 lb
Weight: 17,700 lb
Track area at zero penetration: 11,780 sq in.
Ground pressure, loaded: 2.5 psi
Width, tracks on: 116 in.
Length: 278 in.
Height: 114 in.

Speed: 22 mph
Fording depth: 60 in.
Ground clearance: 17 in.
Gradeability
    forward: 60%
    side: 50%

FIGURE 8.5   Foremost tracked carrier model 30T

Load capacity: 30 tons
Continuous horsepower: 215 hp
Top speed: 17 mph
Track width: 54 in.

Track area: 34,000 sq in.
Fording depth: 66 in.
Deck length: 30 in.
Deck width: 108 in.

(a)

FIGURE 8.6   Two views of a Go-Tract model 200

Weight
    Vehicle: approximately 18,500 lb
    G.V.W.: cross-country 37,000 lb
Length, without winch and bumper: 214 in.
Width: 96 in.
Ground clearance: laden @ G.V.W. 17 in.
Track width: 16 in.
Speed @ maximum engine rpm (governed): 15 mph
Speed @ maximum engine torque: 8 mph
Gradeability
    (a) ascending and descending slope: 60%
    (b) side-slope: greater than 40%
Fording depth: 30 in.

(b)

POWER TRAIN
Engine: 183 bhp @ 3800 rpm
Steering unit: Go-tract controlled differential, ratio 3.27:1
Suspension: trailing arms with hollow rubber springs
Track tensioner:
    Hydraulic rams displace idler bracket to
    continuously and automatically adjust
    track tension
Roadwheels: aluminum with polyurethane tires
Track: space type
    Bolted single pitch construction comprising flexible
    units, special steel guide plates and 16-inch special
    steel grousers

FIGURE 8.7   Nodwell tracked carrier model RN 35

POWER TRAIN:
Engine:
    Ford 240 cu in. gasoline
    net bhp @ 4000 rpm: 120
Crossdrive: Nodwell No. 12 controlled differential
Length over pintle hook: 192¾ in.
Width, over tracks: 105 in.
Ground clearance loaded: 14 in.
Standard track width: 28 in.
Tires: 6.50–16 × 6 pr
Suspension: independent crank arm with torsion spring
Weight: basic unit with cab, deck, stake sides and winch 8,300 lb
Payload: 3,500 lb
Ground pressure
    unloaded: 1.36 psi
    loaded: 1.94 psi
Gradeability
    front: 60%
    side: 30%
Fording depth: 38 in.

FIGURE 8.8   Nodwell tracked carrier model RN 110

POWER TRAIN
Engine: 170 to 234 hp
Speed: 1.8 to 16 mph
Weight: 12,000 net; 23,000 gross
GROUND PRESSURES:
Series 1:
    Unloaded @ zero penetration: 1.1

Loaded @ zero penetration: 2.1
    Loaded @ 10 in sinkage: 1.7
Track assembly width: 40 in.
Track area @ zero penetration: 10,720 sq in.
Width, over tracks: 109 in.
Length, overall: 231¼ in.
Ground clearance, loaded: 16 in.

FIGURE 8.9   Nodwell tracked transporter model RN 400

Ground pressure @ zero penetration
    Unloaded: 1.8 psi
    Loaded: 3.25 psi
Track area: 28,000 sq in.
Ground clearance, loaded: 16 in.
Fording depth: 48 in.
Gradeability: 60%
Maximum speed @ 2500 rpm: 10.8 mph
Weight
    Unloaded (net): 50,000 lb
    Payload (tare): 40,000 lb
    Loaded (gross): 90,000 lb
Length: 544 in.
Width, over tracks: 123 in.
Height: 125½ in.
Deck length: 360 in.
Deck width, 108 in.
Engine (max) bhp: 130 @ 2800 rpm

FIGURE 8.10   Sure-Go tracked carrier

Length: 99 in.
Width: 48 in.
Height: 50 in.
Weight: 1075 lb
Road clearance: 10 in.
Speed: 16 mph
Track pressure: 0.51 psi (tare)
Engine: 16.5 hp

REFERENCES

BEKKER, M. G. 1956. Theory of land locomotion. The University of Michigan Press, Ann Arbor.

BOGDANOFF, J. L., F. KOZIN, and L. J. COTE. 1965–66. Introduction to a statistical theory of land locomotion (in four parts). J. Terramechanics, Vols. 2 and 3.

———— 1966. Atlas of off-road ground roughness P.S.D.S. and reports on data acquisition technique. Land Locomotion Lab., US Army Tank Automotive Center, Tech. Rept. No. 9387 (LL109), Warren, Michigan.

KOZIN, F., L. J. COTE, and J. L. BOGDANOFF. 1963. Statistical studies of stable ground roughness. Land Locomotion Lab., US Army Tank Automotive Center, Tech. Rept. No. 8391 (LL95), Warren, Michigan.

LISTON, R. A. and B. HANAMOTO. 1966. The drawbar pull-weight ratio as a measure of vehicle performance. Land Locomotion Lab., US Army Tank Automotive Center, Tech. Rept. No. 9349 (LL107), Warren, Michigan.

RADFORTH, J. R. 1967. Hybrid computer simulation of terrain-vehicle systems. Proc. Twelfth Muskeg Res. Conf., NRC, ACGR Tech. Memo. 90, pp. 63–69.

RADFORTH, N. W. 1952. Suggested classification of muskeg for the engineer. Eng. J., Vol. 35, No. 11, pp. 1199–1210.

———— 1954. Palaeobotanical method in the prediction of subsurface summer ice conditions in northern organic terrain. Trans. Roy. Soc. Can., Vol. XLVIII, Series III, Section Five, pp. 51–64.

———— 1958. Theory of measurement in relation to drainage and bearing strength of muskeg. Proc. Fourth Muskeg Res. Conf., NRC, ACSSM, Tech. Memo. 54, pp. 59–79.

REECE, A. R. 1964. Problems of soil vehicle mechanics. Land Locomotion Lab., US Army Tank Automotive Center, Tech. Rept. No. 8470 (LL97), Warren, Michigan.

THOMSON, J. G. 1957. A study of some factors influencing vehicle mobility in snow. Defence Research Board, Directorate of Eng. Res., DER Rept. No. 1, Ottawa.

———— 1960. Muskeg and transportation. Can. Pulp and Paper Assoc., Woodlands Section Index No. 1997 (B-8-a).

———— 1961. Vehicle mobility performance in muskeg – a second report. Trans., Can. Inst. Mining Met., Vol. LXIV, pp. 94–100.

U.S. ARMY. 1959. Trafficability of soils. U.S. Army Corps of Engrs. Waterways Exper. Sta. Tech. Memo. No. 3–240 and Suppls. 1–15, Vicksburg, Mississippi.

INDEX

Aapa moor, xvii, 252
Absolute specific gravity. *See* Specific gravity
Access, off-road, 9; frozen muskeg, 244; vehicles, 20, 21, 26, 261–85
Access roads, 27, 181–92. *See also* Roads, secondary and access
Acidity, 31; cause of corrosion, 89, 180, 248; definition, xvii; field measurements, 89; influence of on peat formation, 6; pH values, 82–4, 89, 248; total soil acidity, 180, 249; versus organic content, 118
Aerial surveys
  aerial photogrammetric mapping, 158, 182
  airphoto interpretation, 27
  approach, 53–68, 182
  identification of: airform patterns, 53–9; confined and unconfined muskeg, 66; topography, 65, 243
  prediction of: depth, 65; engineering problems, 66; water factors, 65, 218
  prediction summaries, 66–77
  route selection (roads, pipelines, etc.), 127, 158, 182, 196, 242, 247
  site selection (airstrips), 218
Airform patterns
  definition, xvii
  high altitude, 53–8: dermatoid, *see* Dermatoid; marbloid, *see* Marbloid; reticuloid, *see* Reticuloid; stipploid, *see* Stipploid; terrazoid, *see* Terrazoid
  low altitude, 58–9: apiculoid, *see* Apiculoid; cumuloid, *see* Cumuloid; intrusoid, *see* Intrusoid; planoid, *see* Planoid; polygoid, *see* Polygoid; vermiculoid, *see* Vermiculoid
  map of, 57
  prediction from, 59–68
  use in interpretation, 65
Airstrips
  conventional, 217
  winter, 217–19: categories of, 218; clearing for, 218–19; construction of, 219; site selection, 218
A-line, xvii, 90
Alligator cracks in road surface, 181
Aluminum pipe culverts, 179, 180
Amorphous-granular peat type, 32, 35; compressibility, 107; consistency limits, 90; definition, xvii, 100; permeability, 85; shear strength, 132, 134, 136, 162, 163, 172; structure, 35, 39, 78, 79

Anchors, types of: earth, 226, 238, 240; fluke, 48, 49; grouted, 233, 234, 237; log, 225–6, 227; screw, 225, 227, 232–3, 234, 236
Angle of internal friction, 96–7, 210
Anisotropy of permeability, 85, 95, 138
Apiculoid airform pattern: definition, xvii; description, 59; examples, 61; mapping symbol, 64
Ash content: definition, xvii; determination, 89; relation to shear strength, 96; use in specific gravity determination, 88–9; values, 82–4, 89
ASTM, 32
Atterberg limits. *See* Consistency limits
A-value, xvii, 103
Axon, xvii, 32

Beam test: sphagnum peat, 100
Bearing capacity, 6, 21, 22, 32, 66, 98, 103–6, 159; definition, xvii; effect of peat structure and water content, 263; empirical methods of calculating, 106; formulae, 104, 105, 106; influence on vehicle mobility, 262, 266; punching pressure, 104; safe bearing capacity, 106, 223–4; surface mat, 104, 114; theoretical concepts, 104–5; ultimate bearing capacity, 104, 105, 106
Bearing failure, 264
Bearing stresses, 225
Benkelman beam, 167, 181
Berms, 71, 114; dyke construction, 211, 213, 215; frozen muskeg, 245; railway construction, 206; road construction, 153, 162
Blanket barriers, 262
Blanket bog, xvii, 73, 74
Block sampling, 109, 145, 146
Bog, xvii, 3, 4
Bog blasting, 151, 201; ditching, 247, 258, 259; toe shooting, 165–6, 202; underfill, 165, 202
Bridge abutment, in muskeg, 155
Bridging muskeg, 176, 203–4
British Mires Research Group, 46
Building foundations, 219–21; end bearing piles, 13, 220; excavation and backfill, 220; friction piles, 221; raft foundation, 221; site preloading, 221; use of sand drains, 221